Towards Wireless Heterogeneity in 6G Networks

The connected world paradigm effectuated through the proliferation of mobile devices, Internet of Things (IoT), and the metaverse will offer novel services in the coming years that need anytime, anywhere, high-speed access. The success of this paradigm will highly depend on the ability of the devices to always obtain the optimal network connectivity for an application and on the seamless mobility of the devices. This book will discuss 6G concepts and architectures to support next-generation applications such as IoT, multiband devices, and high-speed mobile applications. IoT applications put forth significant challenges on the network in terms of spectrum utilization, latency, energy efficiency, large number of users, and supporting different application characteristics in terms of reliability, data rate, and latency. While the 5G network development was motivated by the need for larger bandwidth and higher quality of service (QoS), 6G considerations are supporting many users with a wide application requirement, lowering network operating cost, and enhanced network flexibility. Network generations beyond 5G are expected to accommodate massive number of devices with the proliferation of connected devices concept in connected cars, industrial automation, medical devices, and consumer devices.

This book will address the fundamental design consideration for 6G networks and beyond. There are many technical challenges that need to be explored in the next generation of networks, such as increased spectrum utilization, lower latency, higher data rates, accommodating more users, heterogeneous wireless connectivity, distributed algorithms, and device-centric connectivity due to diversified mobile environments and IoT application characteristics. Since 6G is a multidisciplinary topic, this book will primarily focus on aspects of device characteristics, wireless heterogeneity, traffic engineering, device-centric connectivity, and smartness of application.

Dr. Abraham George is the professor and head in the Department of Computer Science and Information Technology at Alliance University, Bangalore, India. He earned his doctorate degree in computer science from the University of Louisville, and master's degree in computer science and communication and information sciences from Ball State University, Indiana.

Dr. George has more than a decade and a half of industry experience with such multinational companies as Kyocera and National Instruments. While in industry, he managed product development teams that created complex and innovative products. Primary areas of interest of Dr. George include wireless networks, distributed computing, and machine learning. His PhD dissertation focused on developing mobility management protocols and algorithms for heterogeneous wireless systems with multi-hop capability that will enable users to obtain the best connectivity. He has several publications in top-rated journals and conferences.

Dr. G. Ramana Murthy is a professor in the Department of Electronics & Communication Engineering at Alliance University, Bangalore, India. His area of expertise is in VLSI and embedded systems. Dr. Murthy holds a doctor of philosophy from Multimedia University, Malaysia. Dr. Murthy has collaborated with various companies including Infineon Technologies, Intel, and MIMOS in Malaysia. He has more than two decades of overseas academic and administrative experience with institutions such as University of Northumbria in Newcastle, the United Kingdom, and Multimedia University, Cyberjaya, Malaysia. His research interests include VLSI, embedded systems, device modeling, memory optimization, low power design, FPGA, and evolutionary algorithms.

Towards Wireless Heterogeneity in 6G Networks

Edited by
Abraham George
G. Ramana Murthy

CRC Press
Taylor & Francis Group
Boca Raton London New York

CRC Press is an imprint of the
Taylor & Francis Group, an **informa** business

Designed cover image: © Shutterstock

First edition published 2024
by CRC Press
2385 NW Executive Center Drive, Suite 320, Boca Raton FL 33431

and by CRC Press
4 Park Square, Milton Park, Abingdon, Oxon, OX14 4RN

CRC Press is an imprint of Taylor & Francis Group, LLC

ISBN: 978-1-032-43830-6 (hbk)
ISBN: 978-1-032-43831-3 (pbk)
ISBN: 978-1-003-36902-8 (ebk)

DOI: 10.1201/9781003369028

Typeset in Sabon
by MPS Limited, Dehradun

Contents

Contributors

Geetha A.
Department of Computer Science &
  Engineering
Alliance University
Bangalore, India

Sunil Chinnadurai
Department of Communication
  Engineering
School of Engineering and Sciences
  (SEAS)
SRM University
Andhra Pradesh, India

Ravuri Daniel
P V P Siddhartha Institute of
  Technology
Vijayawada, Andhra Pradesh, India

Abraham George
Alliance University
Bangalore, Karnataka, India

Rakshit Govind T.
Department of Electronics and
  Communication Engineering
BMS Institute of Technology and
  Management
Yelahanka, Bengaluru, Karnataka,
  India

M. W. Hussain
Department of Computer Science &
  Engineering
Alliance University
Anekal, Karnataka, India

Srikanth Itapu
Department of Electronics and
  Communication Engineering
Alliance University
Bangalore, India

Karan R. Jagdale
Department of Computer Science &
  Engineering
Alliance College of Engineering &
  Design Alliance University
Bangalore, Karnataka, India

Diana Jeba Jingle
Department of Computer Science
  and Engineering
Christ University
Bangalore, India

N. Chitra Kiran
Department of Electrical and
  Electronics Engineering
Alliance University
Bangalore, India

S. Mohan Krishna
Department of Electrical and
  Electronics Engineering
Alliance University
Bangalore, India

Vartika Kulshrestha
Department of Computer Science &
  Engineering
Alliance College of Engineering &
  Design Alliance University
Bangalore, Karnataka, India

Neelapala Anil Kumar
Department of ECE
Alliance University
Bangalore, India

Punam Kumari
Department of Computer Science &
  Engineering
Alliance University
Bangalore, India

Vivek Menon U.
Department of Communication
  Engineering
School of Electronics Engineering
  (SENSE)
VIT University
Vellore, Tamil Nadu, India

Inbarasan Muniraj
Department of Electronics and
  Communication Engineering
Alliance University
Bangalore, Karnataka, India

G. Ramana Murthy
Department of ECE
Alliance University
Bangalore, India

B. Nivetha
Department of Communication
  Engineering
School of Electronics Engineering
  (SENSE)
VIT University
Vellore, Tamil Nadu, India

P. Mano Paul
Department of Computer Science &
  Engineering
Alliance University
Bangalore, India

Shaik Rajak
Department of Communication
  Engineering
School of Engineering and Sciences
  (SEAS)
Andhra Pradesh, India

Vetriveeran Rajamani
Department of Electronics and
  Communication Engineering
Alliance University
Bangalore, Karnataka, India
and
Department of Communication
  Engineering
School of Electronics Engineering
  (SENSE)
VIT
Vellore, Tamil Nadu, India

Poongundran Selvaprabhu
Department of Electronics and
  Communication Engineering
School of Electronics Engineering
  (SENSE)
Vellore Institute of Technology
Vellore, Tamil Nadu, India

R. Shekhar
Department of Computer Science &
    Engineering
Alliance University
Bangalore, India

Sridhar Thota
Department of Electronics and
    Communication Engineering
Alliance College of Engineering and
    Design
Alliance University
Anekal, Bangalore, Karnataka,
    India

Ravi Sekhar Yarrabothu
Department of ECE
VFSTR
Vadlamudi, India

Chapter 1

6G

Opportunities and challenges

Sridhar Thota[1] and Rakshit Govind T[2]

[1]Department of Electronics and Communication Engineering, Alliance College of Engineering and Design, Alliance University, Anekal, Bangalore, Karnataka

[2]Department of Electronics and Communication Engineering, BMS Institute of Technology and Management, Yelahanka, Bengaluru, Karnataka

1.1 INTRODUCTION

1.1.1 Wireless communications

In general, communication systems can be wired or wireless and the communication medium can be guided or unguided. In wired communication, a physical path, such as coaxial cables, twisted pair cables, and optical fiber links, guides the signal as it travels from one point to another; hence, the medium is known as a guided medium. In wireless communication, a physical medium is not necessary, but the signal travels through space instead. The medium utilized in wireless communication is referred to as unguided medium since space alone permits the transmission of signals without any kind of guidance. The transmission of voice and data over a wireless network is free of cables or wires. The most dynamic and rapidly expanding technology area in the communication industry is wireless communication.

In a communication system, information is typically sent across a short distance from a transmitter to a receiver. The transmitter and receiver can be located anywhere between several meters; an example is a TV remote control and a thousand kilometers with the use of wireless communication like satellite communication. Our lives revolve around effective and efficient communication. Wireless communication systems like mobile phones, remote controls, cellular phones, and Bluetooth audio are some of the ones we use most frequently in our daily lives.

1.1.2 Types of wireless communication systems

People require mobile phones today for a variety of purposes, including conversing, using the internet, and multimedia. All these services must be made accessible to the user while they are mobile, or on the go. These wireless communication services allow us to send voice, data, movies, photos, and

DOI: 10.1201/9781003369028-1

other types of content. Other services offered by wireless communication systems include video conferencing, cellular phone, paging, TV, radio, etc.

According to Chowdhury et al. (2020), mobile usage is expected to drastically increase from 2020 to 2030. Mobile subscriptions will increase from 10.7 to 17.1 billion, smartphone usage will increase from 1.3 billion to 5.0 billion, traffic volume from usage will increase 62 to 5,016 EB/month, and traffic per subscriber will increase tremendously from 10.3 to 257.1 GB/month for 6G wireless communication systems.

1.1.3 Generations of wireless communication

There are a lot of improvements in wireless communication and in terms of their capabilities, as per their capabilities and duration of use; they are grouped into different generations. The different generations and their services and capabilities are:

- *0th Generation:*

  - Radio telephones, which is pre-cell mobile technology, communication possible through voice only.
  - These telephones are usually placed in vehicles.

- *1st Generation (1G): 1G was introduced in 1980s.*

  - In mobile communications, first-time calling, using analog signals, was introduced;
  - Frequency division duplexing (FDD) with a bandwidth of 25 MHz introduced.
  - No roaming with small coverage area between operators; and
  - Speed of 2.4 kbps with low sound quality.

- *2nd Generation (2G): 2G was introduced in early 1990s.*

  - This supported the digital shift from analog to digital;
  - Both voice and SMS;
  - Wireless industry with four sectors: digital cellular, mobile data, wireless local area network (WLAN), and personal communication service (PCS) are supported;
  - Minimum to moderate mobile data service;
  - Better coverage area with high data rate provided by 2G WLAN; and
  - Speed of 64 kbps.

The 2.5G introduced the concept called general packet radio services (GPRS).

The 2.75G or enhanced data for global evaluation (EDGE) is faster, with an internet speed of 128 kbps.

- *3rd Generation (3G): 3G was introduced in 2001.*

  - Improved Internet services;
  - Wireless internet services with high speed;
  - Universal mobile telecommunication services (UMTS); and
  - Speed of 2 mbps.

- *4th Generation (4G): 4G was introduced in 2010.*

  - Long-term evaluation (LTE) for the internet services;
  - IP-based protocols;
  - Vo-LTE (voice over LTE) both for voice and the internet;
  - Low transmission cost supports multimedia service;
  - Any desired service, having freedom and flexibility can be selected;
  - Good usability with HD quality streaming; and
  - Speed of 100 mbps.

1.2 5G TECHNOLOGIES

The cellular providers for 5G, fifth generation of broadband mobile networks, have started rolling out internationally since 2019. 5G is the anticipated replacement for the 4G networks that access the majority of present-day cell phones. By 2025, the GSM Association and Statista projects that 5G networks will have more than 1.7 billion subscribers and represent 25% of the global market for mobile technology.

The introduction of 5G technology in India happened in the month of October 2022; however, only a few cities are likely to have 5G services by the year 2024. Since 5G technology is still not available in India, some people would concentrate on it.

Like their predecessors, 5G networks are cellular networks with discrete geographic units called cells serving as the service area. Through a local antenna within the cell, all 5G cellular devices are connected to the internet and phone network via radio waves. With maximum download speeds up to 10 gigabits per second (G bit/s) are available on the new networks.

Since 5G has more capacity and can connect more types of devices, especially busy places, the quality of internet services will improve to a major extent. The networks are anticipated to be used more frequently as internet service providers (ISPs), generally for laptops and desktop computers, competing with current ISPs like cable internet, and will also enable new applications on the Internet of Things (IoT) and machine-to-machine (M2M) spaces. Up to 1 million devices per square kilometer are anticipated for 5G. 3GPP, a standards-setting industry organization, defines "5G" as any system utilizing 5G NR (5G New Radio) software; this definition became widely accepted by late 2018.

1.2.1 Application areas of 5G

According to Yu et al. (2017), three key application areas for the expanded 5G capabilities have been identified by the ITU-R; firstly, massive machine type communications (MMTC), second ultra-reliable low latency communications (URLLC), and finally enhanced mobile broadband (eMBB). Only eMBB will be in use by 2023; but the mMTC and URLLC won't be available until after that.

5G provides quicker connections, more throughputs, and greater capacity than 4G LTE mobile broadband services; enhanced mobile broadband (eMBB) leverages 5G. Greater traffic places like stadiums, cities, and concert venues will profit from this timeline.

Using the network for mission-critical applications, the demand continuous and reliable data interchange is referred to as URLLC. The wireless communication networks use short-packet data transmission to satisfy their latency and reliability needs.

According to the article published by Intel in 2018, a lot of devices will be connected using massive machine-type communications (mMTC). More than 50 billion connected IoT devices will be connected by 5G technology. Most people will opt to use the free Wi-Fi. Drones that broadcast over 4G or 5G will help with disaster recovery operations by giving first responders access to real-time data.

According to Ford, autonomous vehicles will be fully capable of operating without C-V2X. For many services, many cars will have a 4G or 5G cellular connection. Since autonomous vehicles must be able to function without a network connection, vehicles may no longer require 5G. According to the teleoperated driving (ToD) report, the majority of autonomous cars also include teleoperations to complete their missions.

1.2.2 Performance of 5G

1.2.2.1 Speed

Depending on the RF channel and base station (BS) load, 5G speeds will range from about 50 Mbps to 1,000 Mbps (1 Gbps). The mm wavebands would have the fastest 5G speeds, reaching 4 Gbit/s with carrier aggregation and MIMO (assuming a perfect channel and no other BS load).

1.2.2.2 Latency

Depending on the type of handover, latency increases significantly during handovers, ranging from 50 to 500 milliseconds. Research and development efforts are still being made to cut down on handover interruption time.

The optimal "air latency" for 5G is between 8 and 12 milliseconds, omitting retransmissions from hybrid automatic repeat request (HARQ), handovers, etc. For accurate comparisons, "air latency" must be multiplied by backhaul latency to the server and retransmission latency. According to Verizon, the 5G early deployment's latency is 30 ms. The latency can likely be reduced to 10–15 ms by edge servers located near the towers may be 30 ms. Edge servers close to the towers can probably reduce latency to 10–15 ms.

1.2.2.3 Error rate

The bit error rate (BER) is kept to an exceptionally low level in 5G by using an adaptive modulation and coding scheme (MCS). A lower MCS, which is less error prone, will be used by the transmitter once the error rate exceeds a (very low) threshold. Speed is traded in this way to guarantee a nearly error-free system.

1.3 6G TECHNOLOGIES

According to Nayak et al. (2020), many current technologies will be redefined and reorganized during the 6G communication transit period. Existing technology being modified will change our way of life, society, and economy. The term *IoT* in a short span of time will be reinterpreted with another term called "Internet of Everything" (IoE), heralding the emergence of numerous cutting-edge technologies. The intelligent industrial internet of everything (IIIoE), even urban areas are termed *intelligent city*, chain of hospitals as intelligent healthcare, intelligent grid, and intelligent robots will all be made possible by IoE in the 6G connectivity age. Additionally, we anticipate that the proliferation of 6G communication technologies will give rise to numerous new applications.

The panorama of several 6G applications is shown in Figure 1.1.

Using 6G technology, the Internet of Everything (IoE) is made possible, as shown in Figure 1.1. Future reliance on 6G communication technology is shown for intelligent vehicles, drones, transportation, hospitals, agriculture, railways, and ships. In addition to becoming the standard communication protocol for various products and technologies, 6G will no longer be limited to smartphones.

According to Wang et al. (2022), since the global modernization of 5G technology, many countries in Asia like China, USA, and a few European nations have reportedly been investigating 6G technologies for mobile

Figure 1.1 6G landscape of applications.

communication networks. The success of 5G technologies will undoubtedly serve as a foundation for 6G. Because of the advancement of 6G technology, globally we will enter a new era of intelligence starting in the year 2030. The 6G mobile network will incorporate information science and technologies including high-end cellular computing, big data analytics, artificial intelligence (AI), and blockchain in addition to research achievements in domains like mathematics, physics, biotechnology, and materials. The 6G technologies will become the cornerstone of everyone's daily lives, industrial production, and green development as a result of this integration, which will hasten the convergence of different communications with the sensing, processing, and control.

You et al. (2021) predicted that 6G technologies will feature mainly four paradigm shifts: first, one to serve complete global coverage and developing space-air-ground-sea-integrated networks. Second, utilizing complete spectra that is sub-6G Hz, millimeter wave (mm wave) and terahertz (THz) optical frequency bands, etc. to deliver a higher data rate and network capacity. Third, one is exploring machine learning (ML) and finally, the big data and artificial intelligence.

1.3.1 Vision and application areas of 6G technologies

A thousand times greater network capacity would be possible with 6G, along with the rate of terabit-level transmission, the latency of sub-millisecond, centimeter-level location accuracy with capability of devices having trillion connectivity, and a further decrease in overall network energy usage. The *intelligent connection of everything digital twin* society would be realized with the development of a widely interconnected intelligent network under the influence of 6G. With the help of 6G services, several new application scenarios with immersive, intelligent, and omnipresent features will emerge. The 6G network will enable deep connections and interactions for everything, providing the most immersive experience for people, according to the immersive application of 6G mentioned as immersive cloud extended reality (XR), holographic communications, sensory integration, and intelligent interaction the four common use cases.

1.3.1.1 Immersive cloud XR: A broad virtual space

Virtual reality (VR), augmented reality (AR), and mixed reality (MR) are together referred to as XR. Contents are stored, and all rendering and processing tasks are carried out on the cloud in cloud-based XR. This liberates users from the limitations of wires while significantly reducing the computing burden and energy consumption of XR devices. As a result, portable XR gadgets would become commonplace, ensuring more intelligent and immersive experiences while enabling commercialization.

From 2030, massive network and XR terminal improvements will propel XR technology into an age of total immersion. To enable digital transformation in a variety of industries, including commerce and trade, industrial production, culture and entertainment, education and training, and healthcare, cloud-based XR will be connected with the public networks, cloud servers, big data, and AI. Users will be able to engage with environments through voice and motion, including eye, head, and hand gestures, thanks to cloud-based XR. This premium experience can be achieved only when ultralow latency, ultra-reliability, and ultrahigh bandwidth are offered in predictable conditions. This necessitates unique designs at all layers, from physical to link to network.

1.3.1.2 Holographic communications: Extremely immersive experience

Holographic technology will enable three-dimensional (3D) dynamic interactions among people, things, and surrounding environments through natural and lifelike visual restoration, greatly empowering future-oriented communications. This is made possible by the ongoing development of wireless networks, high-resolution rendering, and terminal displays.

The networks must be able to deliver a Tbps level perceived throughput because immersive multidimensional interactions would entail hundreds of concurrent data streams. Loss of information leads to a retransmission for remote microsurgical procedures and holographic therapy, which in turn fails to meet the requirements of reliability and latency. This increases the bar for networks' reliability and security of transmission even higher.

1.3.1.3 Sensory interconnection: Fusion of all senses

We need our senses of sight, hearing, touch, smell, and taste to comprehend the world around us. From 2030, the transmission of signals involving not only the senses of hearing and sight but also those of touch, smell, and taste will play a significant role in communications and be utilized in a variety of industries, including healthcare, education, entertainment, traffic control, production, and social interactions. In the future, even when family members are far apart, people will still be able to experience the warmth of a hug from a loved one through their mobile devices. Users could be able to take advantage of experiences like stunning landscapes and even a stroll on a sandy beach while feeling the sea breeze in the Maldives from the comfort of their own homes.

The coordinated and synchronized transmission of data relating to many senses is necessary for the combined interaction of the major senses. Millisecond-level latency is necessary to maintain a high-quality experience. The need for highly precise positioning is increased by the feedback of the touch sensation, which is directly tied to body movement and location. If networks' maximum capacity is not raised, the in-step transfer of all-sense data will not be feasible. To ensure privacy and prevent infringement, strong data security must be guaranteed when more senses become interconnected during communication. To improve data transfer, new joint and independent encoding and decoding modalities are required because each sense should have its own digital representation.

1.3.1.4 Intelligent interaction: Interactions of feelings and thoughts

Researchers will have the chance to make strides in a variety of fields, including emotional and brain-computer interfaces, thanks to the advent of 6G mobile connectivity. The ability of intelligent agents to see, recognize, and think will create a full-fledged replacement for conventional intelligent equipment. People and intelligent agents will develop user-tool relationships that resemble human interactions with feelings and comprehension. Through conversations and facial expressions, these intelligent agents will be able to discern users' psychological and emotional states, reducing health hazards. Mind-controllable robots will be able to help the disabled overcome physiological challenges in their daily lives and employment while acquiring knowledge and skills by supporting lossless transmission of brain information.

1.3.2 6G applications

Saad et al. (2019) attribute the main contribution of the 6G technologies as a bold, forward-looking application, as shown in Figure 1.2.

6G: Driving Applications

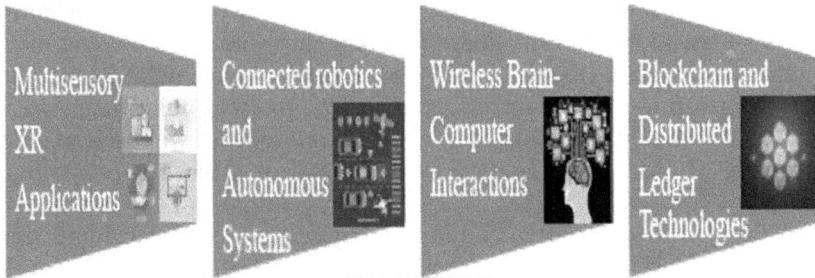

Figure.1.2 Driving applications of 6G.

- *The requirements behind 6G for driving applications:* Traditional applications like live multimedia streaming are the main feature of 6G. The new four prominent application domains are as follows:

1. Multisensory XR Applications;
2. Connected Robotics and Autonomous Systems (CRAS);
3. Wireless Brain-Computer Interactions (BCI); and
4. Blockchain and Distributed Ledger Technologies (DLT).

Transition from smart to intelligent is prominent in 6G technologies; in this case the "Quality of Services" is an important factor. They are:

1. 6G Enabling Technologies
   A. Internet of Everything (IoE)
   B. Edge Intelligence
   C. Artificial Intelligence

2. 6G Vehicular Technologies
   A. Intelligent Cars
   B. Unmanned Aerial Vehicles
   C. Intelligent Transportation

- *Spectrum bands:* Spectrum bands are a key component in enabling radio connection. Every new mobile generation needs a new pioneer spectrum that aids in maximizing the advantages of a cutting-edge technology. It will also be crucial to transition the current mobile communication spectrum from the older technology to the new one. The new pioneer spectrum blocks for 6G are anticipated to be at low bands (460–694 MHz) for extreme coverage and sub-THz for peak data rates exceeding 100 Gbps, and mid-bands (7–20 GHz) for urban outdoor cells enabling larger capacity through extreme MIMO.

A network that can sense the ability of 6G to perceive the environment, people, and objects would be its most prominent feature. By collecting signals that are reflecting off objects and identifying their type, shape, relative location, velocity, and possibly even material qualities, the network transforms into a source of situational information. In conjunction with other sensing modalities, such a form of sensing can assist build a "mirror" or digital counterpart of the physical world, extending our sensations to every point the network touches. By combining this data with AI/ML, the network will become more intelligent and offer fresh perspectives from the physical world.

- *Extreme connectivity:* In order to meet the needs, including sub-millisecond latency, the ultra-reliable low-latency communication (URLLC) service, which debuted with 5G, will be improved and refined in 6G. Simultaneous transmission, several wireless hops, device-to-device connections, and AI/ML could all increase network stability.

The experience of real-time video communications, holographic experiences, or even digital twin models updated in real time through the deployment of video sensors will be improved by improved mobile broadband coupled with decreased latency and greater reliability.

In the 6G era, we can expect use cases with networks that have specific requirements in sub-networks, creating networks of networks with networks as an endpoint. Machine area networks such as a car area network or a body area network can have hundreds of sensors over an area of less than 100 meters. These sensors will need to communicate within 100 microseconds with extreme high reliability for the operation of that machine system. Making networks within cars or on robots, truly wireless will open a new era for the designers of those devices as they would no longer need to install lengthy and bulky cable systems.

- *New network architectures:* The first system intended to function in a business or industrial setting and replace wired connectivity is 5G. Industries will need even more sophisticated designs that can allow greater flexibility and specialization as demand and strain on the network increase.

The core of 6G will feature services-based architecture, and cloud native deployments will be expanded to include portions of the radio access network (RAN). The network will also be set up in a heterogeneous cloud environment that combines private, public, and hybrid clouds. There will also be chances to cut costs by combining functions as the core of the RAN gets more decentralized and the higher layers become more centralized. An unprecedented level of network automation that lowers operational costs will be achieved by new network and service orchestration solutions that take advantage of AI/ML advancements.

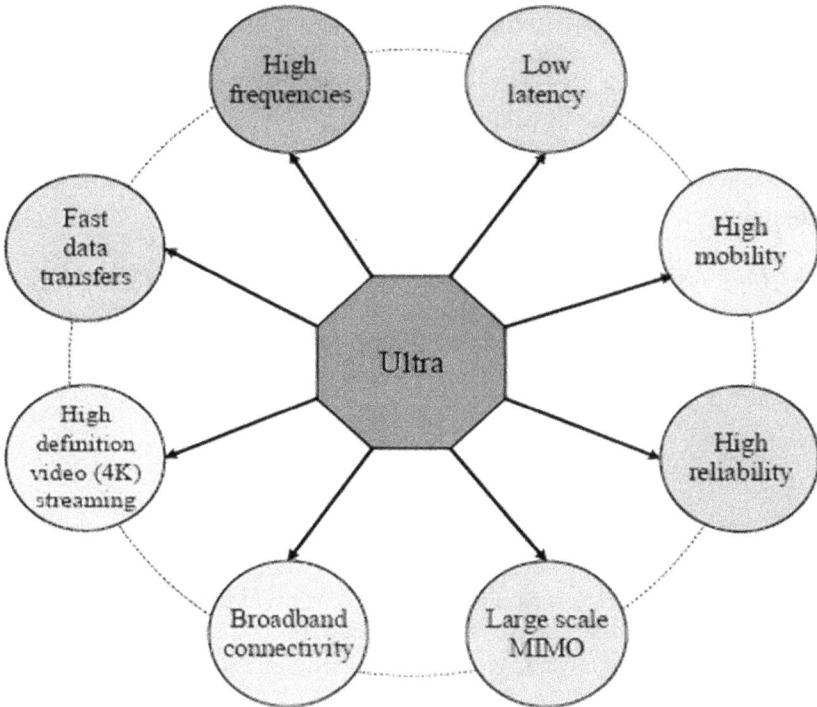

Figure 1.3 Ultra era in 6G networks.

- *Security and trust:* Networks of all kinds are increasingly targets of cyber-attacks in terms of security and trust. Strong security methods must be implemented due to the threats' dynamic nature. The 6G networks will be made to be resistant to dangers like jamming. When developing new mixed-reality environments that include digital representations of actual and imaginary objects, privacy concerns must be considered.

Figure 1.3 shows the ultra era of 6G technologies.

1.4 MOVING TOWARDS INDUSTRY 5.0

Several technology advancements over the past few decades have fuelled the development of smart factories. However, connectivity has remained a significant problem. The fourth industrial revolution was launched by 5G with a plethora of cutting-edge technology. The broad adoption of 6G will give the march towards Industry 5.0 more momentum.

Co-designing communication and control will lead to lower costs, greater data rates, and more use cases. By enabling collaborative communication, sensing, and localization using a single system, the 6G network will serve as a sensor that will lower costs.

Using backscatter communications, new zero energy or battery-less devices might be made possible on the 6G network, enabling a tremendous scale of data collection for analytics and closed-loop control. Mobile robot swarms and drones will be widely used in a variety of industries, including hospitality, healthcare, warehouses, and package delivery.

- *Upgrading to 6G:* The introduction of 5G and then 5G-Advanced could not have occurred at a more advantageous time when global resources are already at a premium. Communications technology will be essential to increasing productivity and pursuing comprehensive environmental policy. By enhancing human well-being and revealing new possibilities that we cannot currently define nor conceive, 6G will further build on the accomplishments of 5G.

According to Nokia, 6G systems will go on sale by 2030, with a normal 10-year gap between generations. Phase 1 of standardization, which is a component of 3GPP Release 20, is likely to begin in 2026.

Although 6G won't arrive until the end of the decade, 5G will be improved by 5G-Advanced, which will be a major 3GPP priority starting with Release 18 and power commercial networks starting in 2025.

Nokia has been a pioneer in establishing the core technologies for the 5G era and beyond, powered by internationally acclaimed research from Nokia Bell Labs. Hexa-X, the 6G flagship initiative of the European Commission for research into the following generation of wireless networks, is being led by Nokia in an effort to make 6G a reality before 2030. Nokia is a founding member of the Next G Alliance, an initiative to advance North American mobile technology leadership, as well as RINGS, an NSF-led initiative in the United States that will speed up research in areas with the potential to have a significant impact on next-generation (NextG) networking and computing systems. These initiatives are in addition to the numerous 6G research projects being conducted around the world.

We currently lack 1 Tbps data speeds, which would be incredibly dependable and low latency, and thus causes QoS to be compromised. Several research opportunities and possibilities exist in accordance with the requirements and promises of 6G communications, which will lead to the creation of several new applications. Thus, the global economy, society, and lives will all benefit from 6G. Many current technologies will be redefined and reorganized in the 6G communication era. Redesigning current technology will change our way of life, culture, and economy. The Internet of Everything (IoE), which will herald the beginning of numerous cutting-edge technologies, will soon replace the Internet of Things (IoT).

IoE will play a major role in the 6G communication age and will enable intelligent cities, intelligent grids, intelligent robots, and intelligent

6G: Driving Trends

6G: Enabling Technologies

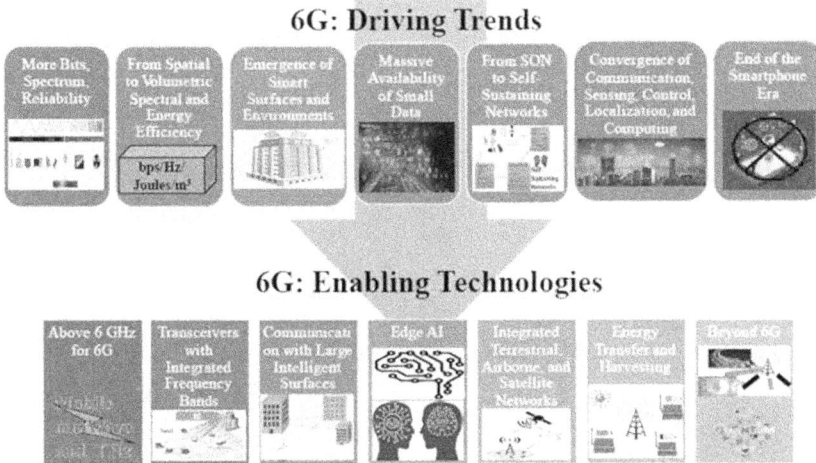

Figure 1.4 Trends and technologies.

healthcare (including intelligent wearable and the intelligent internet of medical things). Additionally, we anticipate that the proliferation of 6G technology will give rise to a wide variety of new applications. As a result, we consider how 6G communication technology may affect numerous fields in this article. In this chapter, the main aspects of 6G technology are summarized.

Additionally, this chapter also investigated the 6G communications' enabling technology. The effects of 6G communications technology on vehicular technology, robots, virtual reality, healthcare, and cities are exposed. It also explores how 6G communication technology can help spark an industrial revolution in trends and enabling technologies, as shown in Figure 1.4.

1.5 CHALLENGES AND FUTURE RESEARCH DIRECTIONS

Several technical problems need to be solved to successfully deploy 6G communication systems. A few possible concerns are briefly discussed below and shown in Figure 1.5.

- *High THz propagation and air absorption:* High data rates are made possible by high THz frequencies. However, because of the high propagation loss and atmospheric absorption characteristics of the THz bands, there is a significant obstacle to data transfer over relatively long distances. For the THz communication systems, we need a novel transceiver architecture design. We need to make sure that extremely broadly available bandwidths are used to their full potential and that the transceiver can work at high frequencies.

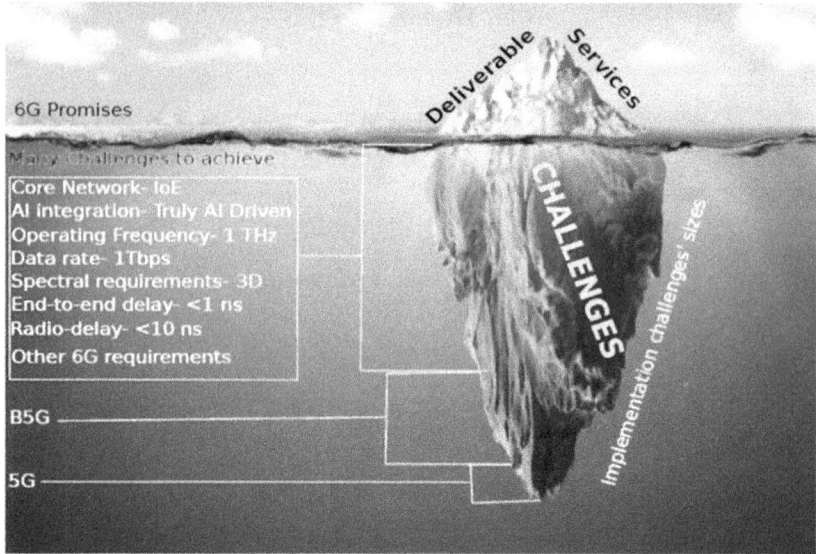

6G Promises

Many challenges to achieve

Core Network- IoE
AI integration- Truly AI Driven
Operating Frequency- 1 THz
Data rate- 1Tbps
Spectral requirements- 3D
End-to-end delay- <1 ns
Radio-delay- <10 ns
Other 6G requirements

B5G

5G

Deliverable Services

CHALLENGES

Implementation challenges' sizes

Figure 1.5 Different challenges of 6G.

Another difficulty with THz communication is the very low gain and effective area of the various THz band antennas. Concerns about THz band communications' impact on public health and safety must also be addressed.

- *Complexity in resource management for 3D networking:* Resource management for 3D networking is complicated since the networking technology was extended vertically. As a result, a new dimension was created. In addition, numerous attackers could intercept reliable data, which could seriously harm system performance. New methods of resource management and optimization for routing protocol, multiple access, and mobility assistance are therefore crucial. A new network structure is required for scheduling.

- *Heterogeneous hardware constraints:* The 6G standard will entail a very large number of heterogeneous communication system types, including frequency bands, communication topologies, service delivery, and more. Additionally, the hardware configurations of the access points and mobile terminals will range greatly. From 5G to 6G, the massive MIMO technique will be significantly improved, which may call for a more complicated design. Additionally, the communication protocol and algorithm design will become more difficult. But communication will also use AI and machine learning. Additionally, the hardware architecture of various communication systems varies. Hardware implementation may get more complicated because of unsupervised and reinforcement learning. As a result, it will be difficult to combine the communication platforms into one.

- *Autonomous wireless systems:* The 6G network will fully support AI-based Industry 4.0, autonomous vehicles, and unmanned aerial vehicles. The convergence of numerous heterogeneous sub-systems, including autonomous computing, interoperable processes, system of systems, machine learning, autonomous cloud, machines of systems, and heterogeneous wireless systems, is required to create autonomous wireless systems. As a result, developing the entire system becomes difficult and complex. For instance, designing a completely automated self-driving vehicle that performs better than human-controlled vehicles will make establishing a fully autonomous system for the driverless vehicle much more difficult.

- *Modeling of sub-mm wave (THz) frequencies:* Because air conditions have an impact on the propagation properties of mm wave and sub-mm wave (THz), absorptive and dispersive effects are observed. The weather is highly unpredictable because it changes regularly. As a result, this band's channel modeling is somewhat complicated, and there is no perfect channel model for this band.

- *Device capability:* Several additional functions will be offered by the 6G system. Smartphones and other devices should be able to handle the new functionalities. Supporting 1 Tbps speed, AI, XR, and integrated sensing with communication features utilizing separate devices is particularly difficult. Some 6G features may not be supported by 5G devices, and the cost of 6G devices may rise as their capabilities improve. We must ensure that the billions of gadgets that will be connected to the 5G technology are also compatible with the 6G technology.

- *High-capacity backhaul connectivity:* Access networks in 6G will feature a very high density and high-capacity backhaul connectivity. Additionally, the form of these access networks is varied, and they are widely dispersed across a region. Each of these access networks will provide connectivity at very high data rates for a variety of user categories. To provide high-data-rate services at the user level, the backhaul networks in 6G must be able to manage the enormous volume of data required for connecting between the access networks and the core network; otherwise, a bottleneck would be created. A challenge for the exponentially increasing data demands of 6G is any increase in the capacity of the optical fiber and FSO networks, which are potential solutions for high-capacity backhaul connectivity.

- *Spectrum and interference management:* Management of the 6G spectrum, including spectrum-sharing schemes and cutting-edge spectrum management systems, is crucial due to the limited availability of spectrum resources and interference problems. To maximize QoS and achieve the highest resource utilization, effective spectrum management is crucial. Researchers working on 6G must handle issues including spectrum sharing and spectrum management in heterogeneous networks that synchronize transmission at the same frequency. The typical interference cancellation techniques, such as parallel interference cancellation and consecutive interference cancellation, need to be investigated further by researchers.

- **Beam management in THz communications:** Beamforming using massive MIMO systems is a technology that has the potential to facilitate high-data-rate communications. However, due to the propagation properties of the sub-mm wave, or the THz band, beam management is difficult. Therefore, effective beam control in the face of adverse propagation characteristics will be difficult for huge MIMO systems. In high-speed vehicular systems, it's crucial to select the ideal beam quickly in order to provide a seamless handover.

Successive and fascinating features are added to communication systems with each successive iteration. Exciting features can be found in the 5G communication system, which will formally launch globally in 2020. But by 2030, 5G won't be able to keep up with the expanding demand for wireless communication. As a result, 6G will have to be implemented. The study stage of 6G research is still in its infancy.

In this study, the possibilities, and strategies for achieving the 6G communication goal are envisioned. In this study, we discussed the potential uses and the 6G communication technologies that would be implemented. To achieve the objectives for 6G, we also discussed potential obstacles and future research directions. We have outlined the various technologies that could be employed for 6G communication in addition to outlining the vision and objectives of 6G communications.

1.6 CONCLUSION

In the wireless communication era, from first generation to date, there are a lot of challenges in front of academia and industry. To date, there is evidence that every decade there is a mobile new generation. Accordingly, the 4G is in place globally and it is expected that 5G will be implemented and it may serve from 2020 to 2030. The research on 6G is in big flow and it may serve from 2030 to 2040, and 7G and so on; hence, in every decade there is a possibility of a new generation. The current 5G is deployed in many countries. Already, 5G communication technology is showing an impact on the global economy. It is predicted that the ongoing generation of 5G will have more impact on the global economy. It is anticipated that the requirements of 6G are more different from 5G communications. Globally, all countries are ready to accept 6G, since the parameters of 6G can support new technologies and applications, which are shown in Figure 1.1. For example, 6G will be a major player in vehicular technology, healthcare, and commercial enterprise. Additionally, 6G communication technology will have an impact on a lot of academics, professionals, and businesspeople. As a result, various study articles have previously covered 6G parameters. Even though global deployment of 6G technology is anticipated starting in 2030, there are many problems and difficulties that arise from its claims. Rural technology adoption is also a significant obstacle. There are problems with mobility and coverage because 5G communication technology is currently geared towards campus solutions. As a result, 5G cannot accommodate a wide range of applications. However, the 6G communication technology promises to use satellite communication to expand coverage and mobility.

REFERENCES

Chowdhury, Mostafa, Shahjalal, Md, Ahmed, Shakil, & Jang, Yeong Min. (2020). 6G Wireless Communication Systems: Applications, Requirements, Technologies, Challenges, and Research Directions. *IEEE Open Journal of the Communications Society*. 957–975. 10.1109/OJCOMS.2020.3010270.

5GAA Tele-Operated Driving (ToD): Use Cases and Technical Requirements Technical Requirements. (https://venturebeat.com/2020/08/17/smooth-teleoperator-the-rise-of-the-remote-controller/)

Nayak, Sabuzima & Patgiri, Ripon. (2020). 6G Communication: Envisioning the Key Issues and Challenges. *EAI Endorsed Transactions on Internet of Things*. 6, 166959. 10.4108/eai.11-11-2020.166959.

Saad, Walid, Bennis, Mehdi, & Chen, Mingzhe. (2019). A Vision of 6G Wireless Systems: Applications, Trends, Technologies, and Open Research Problems. *IEEE Network*. 34(3), 134–142.

Wang, Z., Du, Y., Wei, K. et al. (2022). Vision, Application Scenarios, and Key Technology Trends for 6G Mobile Communications. *Science China Information Sciences*. 65, 151301. 10.1007/s11432-021-3351-5

Yu, Heejung, Lee, Howon, & Jeon, Hongbeom. (October 2017). What Is 5G? Emerging 5G Mobile Services and Network Requirements. *Sustainability*. 9(10), 1848. 10.3390/su9101848 (https://doi.org/10.3390%2Fsu9101848)

You, X.H., Wang, C.X., Huang, J. et al. (2021). Towards 6G Wireless Communication Networks: Vision, Enabling Technologies, and New Paradigm Shifts. *Science China Information Sciences*, 64, 110301

Chapter 2

Disruptive technology directions for 6G

Vartika Kulshrestha and Karan R. Jagdale

Department of Computer Science & Engineering, Alliance College of
Engineering & Design Alliance University, Bangalore, Karnataka, India

2.1 INTRODUCTION

Disruptive technology often starts as niche technology, serving a small
segment of the market with a unique value proposition. Over time, it gains
momentum and improves its capabilities, becoming a viable alternative to
the existing technology or product. Disruptive technology is typically
characterized by its ability to provide more value at a lower cost or to
deliver a new value proposition that the existing technology cannot. The
adoption of disruptive technology can be slow at first, but once it gains
momentum, it can quickly reshape the industry landscape and create new
winners and losers in the market.

As the next frontier in wireless communication, 6G, or sixth-generation
technology, is predicted to bring revolutionary changes to our digital
landscape. While the world is still navigating the initial stages of 5G adoption,
the advent of 6G looms on the horizon (Pouttu et al., 2020), with experts
estimating its arrival by 2030. Table 2.1 provides a comparison between the
network generations. Once fully implemented, 6G aims to deliver astounding
data transfer speeds of up to 100 Gbps (gigabits per second), outpacing 5G's
maximum speeds by a factor of ten or more (Vreman & Maggio, 2019).

A key advancement of 6G is the exploration of the terahertz (THz)
spectrum, which lies between the microwave and infrared range (Khalid
et al., 2021). Tapping into this spectrum allows for groundbreaking data
transfer rates and ultra-low latency. Additionally, 6G will employ sophisticated AI-driven networking strategies, empowering rapid decision-making
and optimal network performance (Nawaz et al., 2021). Another notable
feature of 6G is the integration of advanced sensing technologies, paving
the way for smart systems that adapt to diverse user requirements and
environmental conditions (Saad et al., 2021). 6G technology is anticipated
to have far-reaching economic implications, potentially contributing up to
$17 trillion to the global economy by 2035, according to a recent study
by Qualcomm (2021). The innovation spurred by 6G is poised to create
new industries, streamline existing sectors, and drive technological

DOI: 10.1201/9781003369028-2

Table 2.1 Comparison between 3G, 4G, 5G, and 6G

Case	3G	4G	5G	6G
Peak Data Rate	2 Mbps	1 Gbps	10 Gbps	1 Tbps
Technology	WCDMA	LTE	MIMO	OFDM
Frequency	1.6–2.0 GHz	2.0–8.0 GHz	3.0–30.0 GHz	95 GHz-3.0 THz
Latency	500 ms	100 ms	10 ms	1 ms
Spectral Speed	1 bps/Hz	15 bps/Hz	30 bps/Hz	100 bps/Hz

breakthroughs. Consequently, the 6G era is expected to generate millions of jobs, encourage entrepreneurship, and promote healthy competition among nations (Qualcomm, 2021).

The societal benefits of 6G are multifaceted, affecting areas as diverse as urban development, healthcare, and entertainment. As 6G supports (Akhtar et al., 2020) massive-scale IoT deployment, the concept of smart cities will become increasingly tangible, optimizing energy usage, enhancing public safety, and revolutionizing transportation. In the healthcare domain, 6G will enable remote monitoring, telemedicine, and prompt health data analysis, ultimately facilitating improved patient care and more efficient medical practices. Furthermore, 6G's ultra-low latency will pave the way for truly immersive AR/VR experiences, transforming education, entertainment, and even travel (Figure 2.1).

Identifying disruptive technology directions for 6G is crucial to unlocking the full potential of the next-generation wireless communication system. By anticipating and understanding emerging technological trends, stakeholders can ensure that 6G delivers groundbreaking capabilities, transforming the way we interact with technology. According to a study by Nokia Bell Labs, 6G's impact is expected to be even more profound than that of its predecessors, making it imperative for researchers, policymakers, and industry leaders to

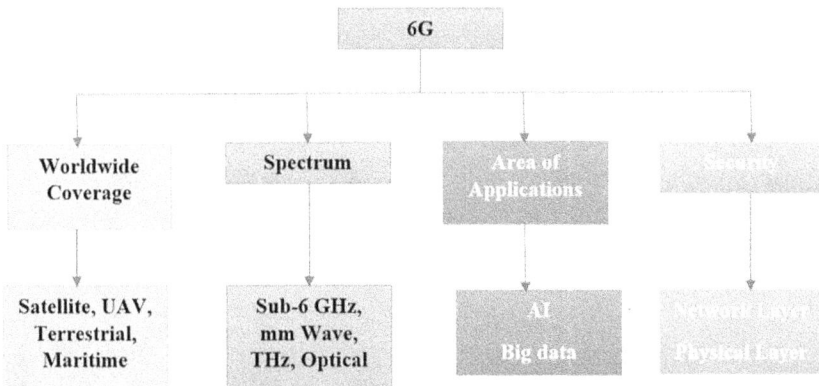

Figure 2.1 6G communication network.

stay ahead of the curve and identify innovative technologies to drive 6G development.

The importance of identifying disruptive technology for 6G goes beyond mere technological progress, as it has far-reaching implications for society at large. Harnessing innovative technology is essential for achieving sustainability goals, including the United Nations' 2030 Agenda for Sustainable Development. As 6G networks incorporate emerging technologies such as holographic communication, quantum computing, and energy harvesting, we can expect remarkable advances in areas such as environmental monitoring, urban planning, and resource optimization, ultimately contributing to a more sustainable and equitable world.

The period between 2020 and 2023 witnessed remarkable progress in the mobile technology landscape, as the transition from 4G to 5G continued to gain momentum worldwide. In 2020, 5G subscribers reached approximately 229 million, representing around 2.9% of total mobile subscribers (GSMA Intelligence, 2021). The rapid expansion of 5G infrastructure and the widespread adoption of 5G-enabled devices spurred significant growth in the number of subscribers, with estimates placing the figure at over 1 billion by the end of 2023 (Ericsson, 2021). During this timeframe, 4G subscriptions remained prevalent, with a gradual decline in the number of 3G subscribers as network providers focused on the deployment of newer technologies. While 6G technology is still in its research and development phase, with commercial availability not expected until around 2030, the rapid increase in 5G subscribers and its transformative impact are paving the way for the eventual introduction of 6G, setting the stage for another leap in connectivity, performance, and global adoption (Figure 2.2).

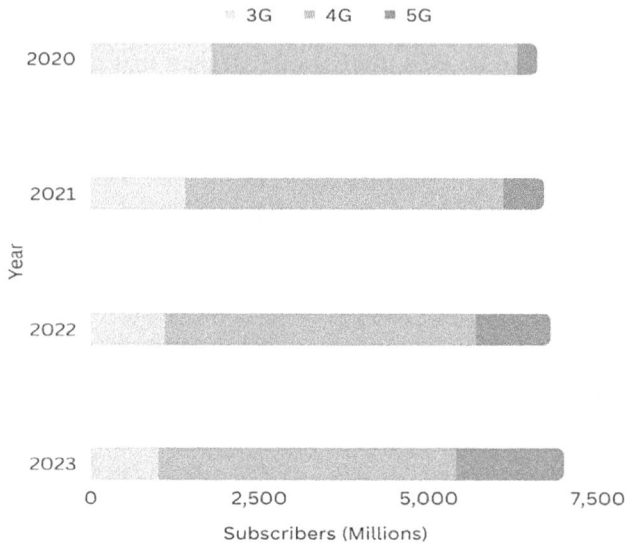

Figure 2.2 Number of subscribers of 3G, 4G, and 5G in the years 2020, 2021, 2022, and 2023.

The integration of artificial intelligence (AI) and machine learning (ML) into 6G networks will be crucial for optimizing network performance and delivering highly personalized user experiences. AI-driven algorithms will play a pivotal role in network management, enabling the efficient allocation of resources and automated optimization in response to fluctuating network conditions. Machine learning techniques can be employed to predict and mitigate potential bottlenecks or outages, proactively addressing issues before they impact the end user. Furthermore, the utilization of AI and ML will help support advanced use cases, such as massive IoT deployments, mission-critical applications, and immersive experiences, by adapting network performance to cater to specific needs and requirements.

The combination of AI, ML, and 6G will pave the way for the development of smart applications across various sectors, including healthcare, transportation, and environmental monitoring. For example, in healthcare, AI-driven systems powered by 6G can enable remote patient monitoring, real-time diagnostics, and precision medicine, significantly improving patient outcomes, and reducing healthcare costs. In transportation, AI and ML can be employed to develop intelligent traffic management systems, support autonomous vehicles, and enhance public transit efficiency. Meanwhile, in environmental monitoring, AI-assisted sensor networks can provide real-time data on pollution levels, natural resource management, and disaster prevention, allowing for more effective decision-making and improved environmental sustainability.

The importance of 6G in supporting next-generation applications cannot be overstated, as it will lay the foundation for a plethora of transformative use cases across various industries. The unprecedented capabilities of 6G, including ultra-high data transfer rates, ultra-low latency, and extreme reliability and reliability, are essential for realizing the full potential of applications like virtual reality, autonomous vehicles, and smart cities. 6G will enable immersive experiences by providing seamless and lag-free connectivity, thus ensuring real-time interaction and high-quality content delivery. In the realm of autonomous vehicles, 6G will facilitate robust vehicle-to-everything (V2X) communication, ensuring the safety and efficiency of self-driving systems. 6G will be instrumental in creating interconnected urban environments for smart cities by integrating IoT devices and smart infrastructure, enabling better resource management, reduced energy consumption, and improved quality of life. In essence, 6G will act as a catalyst for the digital transformation of various industries, paving the way for new opportunities and innovation.

2.2 6G'S SUPPORT FOR EMERGING APPLICATION

2.2.1 Virtual reality

6G technology will play a crucial role in supporting virtual reality (VR) experiences by providing seamless and robust connectivity essential for real-time interactions and high-quality content delivery. The ultra-high data transfer rates and ultra-low latency offered by 6G will enable immersive and interactive VR experiences that were previously impossible due to the limitations of existing wireless technologies. 6G will allow for large-scale multi-user VR environments, where numerous users can simultaneously engage in highly realistic experiences, facilitating improved collaboration, communication, and content sharing. By enhancing the capabilities of VR technology, 6G has the potential to revolutionize various industries, including gaming, education, healthcare, and entertainment, and open new possibilities for innovation and growth.

For any organization to offer this type of immersive technology, 6G networks will need to fulfill several critical requirements. First, ultra-high data transfer rates will be essential to accommodate the enormous amount of data generated by high-quality VR content, including high-resolution video, complex 3D models, and detailed environments. Second, ultra-low latency will be crucial for ensuring real-time interactions, minimizing motion sickness, and maintaining user immersion in VR experiences. This will require network response times of less than 1 millisecond, which is significantly faster than the latency levels achieved by current 5G networks. Additionally, 6G networks will need to provide highly reliable and consistent connectivity to avoid disruptions or performance drops, which can significantly impact the user experience in VR applications.

2.2.2 Autonomous vehicles

Autonomous systems, including self-driving cars, drones, and robots, rely on a continuous, real-time data exchange with their environment to make precise and efficient decisions. By offering a robust and responsive network, 6G will facilitate this critical data exchange, ensuring that autonomous systems have up-to-date information about traffic conditions, road infrastructure, weather, and other relevant factors. Furthermore, the ability of 6G to handle massive numbers of connected devices simultaneously will be essential in accommodating the expected surge in autonomous vehicles, robots, and Internet of Things (IoT) devices in the coming years.

In addition to supporting autonomous systems, 6G is expected to significantly enhance vehicle-to-everything (V2X) communication. V2X communication encompasses the exchange of information between vehicles, infrastructure, and other connected entities, improving traffic management, reducing accidents, and facilitating more efficient transportation systems. With 6G's ultra-reliable low-latency communication (URLLC) capabilities and its support for high data rates, vehicles will be able to communicate seamlessly with each other, as well as with traffic lights, sensors, and pedestrians (Agiwal et al., 2016).

The advancement of V2X communication through 6G will enable vehicles to anticipate and react to potential hazards more effectively. By exchanging real-time information about road conditions and obstacles, vehicles will be better equipped to avoid collisions and adjust their routes accordingly. This enhanced situational awareness will ultimately contribute to safer roadways and more reliable transportation systems. Moreover, 6G-enabled V2X communication will support the coordination of vehicles to optimize traffic flow, reducing congestion and improving overall travel times. 6G networks will facilitate real-time updates about road conditions, traffic, and weather, allowing vehicles to make informed decisions about route planning and driving strategies. This capability will be particularly useful for autonomous systems, which rely on accurate, up-to-date information to navigate their environment safely and efficiently. By integrating advanced sensing technologies and AI-driven networking strategies, 6G will empower autonomous vehicles and other connected systems to make rapid, informed decisions based on a comprehensive understanding of their surroundings, ultimately leading to safer and more efficient transportation systems.

2.2.3 Smart cities

According to a report by the United Nations, 68% of the world's population is projected to live in urban areas by 2050, highlighting the importance of developing sustainable, efficient, and interconnected urban environments. One area where 6G technology is expected to have a significant impact is the integration of IoT devices in smart cities. With the global IoT market predicted to reach $1.1 trillion by 2026, there will be an increasing need for robust and reliable communication networks to support the vast number of connected devices (Fortune Business Insights, 2019). 6G's ability to accommodate massive numbers of connections per unit area, up to 10 million devices per square kilometer, will enable cities to efficiently manage the data generated by IoT devices and extract valuable insights for improved decision-making.

In addition, 6G will be instrumental in the rollout of smart infrastructure like IoT, connected utilities, and public safety networks. 6G's high data rates and low latency will be essential to the real-time monitoring, analysis, and control of these cutting-edge systems. By using real-time traffic data, connected traffic lights, for instance, can optimize traffic flow, drastically cutting down on both congestion and fuel costs. Therefore, cities that invest in 6G-powered smart infrastructure will see substantial cost savings and improvements in sustainability.

6G's role in creating interconnected urban environments extends to facilitating seamless communication between different city systems and services. By enabling real-time data exchange between various sectors, such as energy, transportation, and public safety, 6G networks will empower cities to optimize resource allocation, enhance sustainability, and improve the quality of life for their residents. For instance, smart grids can leverage

6G-enabled IoT devices to manage energy distribution more effectively, contributing to a reduction in energy consumption and greenhouse gas emissions. Ultimately, the integration of 6G technology in smart cities will pave the way for a more connected, efficient, and sustainable urban future.

2.3 CHALLENGES FOR SUPPORTING NEW APPLICATIONS

With each generation of wireless technology offering more capacity and speed to support an expanding range of applications and services, bandwidth requirements have changed significantly over time. It is anticipated that the network capacity of 6G, the next generation of wireless communication systems, will significantly improve in order to meet the growing demands of emerging technologies like virtual reality, autonomous vehicles, and smart cities (Boccardi et al., 2021).

According to a report from Ericsson, the number of cellular IoT connections is expected to increase from 1.3 billion in 2020 to 5 billion by 2026, representing a compound annual growth rate (CAGR) of 25%. This growth is driven by the increasing adoption of IoT devices in various industries, such as healthcare, transportation, and manufacturing (Ericsson, 2021). Moreover, the demand for high-speed data transmission is increasing rapidly. A report from Markets and Markets predicts that the global 5G services market size will reach $41.5 billion by 2026, growing at a CAGR of 29.4% from 2020 to 2026. This growth is attributed to the increasing demand for high-speed internet connectivity, rising mobile data traffic, and the growing adoption of IoT devices.

To meet the growing demand for high-speed data transmission and connectivity, 6G wireless communication systems are being developed with peak data rates of up to 1 Tbps, which is 1,000 times faster than the peak data rate of 5G. According to a report from Global Market Insights, the 6G technology market size is projected to exceed $1 trillion by 2030, growing at a CAGR of over 100% from 2025 to 2030. This exponential growth is attributed to the increasing adoption of 6G technology in various industries, such as healthcare, automotive, and industrial automation. In addition to peak data rates, 6G wireless communication systems are expected to offer extremely low latency, which is crucial for applications that require almost instantaneous response times. According to a report from Research and Markets, the 6G technology market is expected to grow at a CAGR of 140% from 2020 to 2026, driven by the increasing demand for low-latency communication and the growing adoption of AI and machine learning in wireless communication systems.

To achieve these high data rates and low latency, 6G wireless communication systems will utilize advanced technologies such as terahertz (THz) spectrum, massive MIMO, and beamforming. According to a report from Allied Market Research, the global terahertz (THz) technology market size is expected to reach $1.3 billion by 2027, growing at a CAGR of 24.4% from 2020 to 2027. This growth is driven by the increasing adoption of THz technology in various applications, such as medical imaging, security screening, and wireless communication.

Wireless data rates have undergone a dramatic evolution; initially, 2G networks offered speeds of about 200 kbps, while 3G networks offered speeds of up to 2 Mbps. Data rates for 4G technology then exceeded 100 Mbps, and current 5G technology is able to deliver gigabit speeds. The demand for even higher data rates is increasing as the world becomes more connected and data hungry. In comparison to previous generations, 6G is predicted to deliver peak data rates of up to 1 Tbps, enabling unprecedented levels of connectivity and supporting data-intensive applications (Nawaz et al., 2021).

The exploration of new spectrum bands, the use of cutting-edge antenna technologies, and the incorporation of artificial intelligence (AI) and machine learning are just a few of the numerous 6G network capacity-boosting strategies (ML). 6G networks can achieve faster data transfer rates and extremely low latency by utilizing the terahertz (THz) spectrum, which uses frequencies between 100 GHz and 10 THz. This enables instantaneous communication and real-time data processing (Boccardi et al., 2021). Advanced antenna technologies can also increase spectral efficiency and network capacity, enabling more dependable and effective communication. Examples include massive MIMO (multiple-input multiple-output) with hundreds or even thousands of antennas, and beamforming.

The complexity and size of 6G networks will be managed in large part by AI and ML. AI-driven networking strategies can boost network capacity while reducing latency and energy use through intelligent network optimization and resource allocation. According to a recent study, 6G networks could outperform 5G systems in terms of energy efficiency by up to 100 times (Nawaz et al., 2021). As a result, 6G technology is ready to transform wireless communication and open the door to cutting-edge services and applications that were previously unthinkable.

Latency and reliability are two fundamental factors that are expected to have a significant impact on the effectiveness of 6G network performance. The next generation of wireless communication will be tasked with the responsibility of supporting a wide variety of time-sensitive and mission-critical applications. These applications could range from autonomous transportation systems to real-time remote surgeries. In order to ensure that these new applications run without a hitch, it will be necessary to have connectivity that is both extremely dependable and extremely low in latency, which is exactly what 6G intends to provide.

When it comes to applications designed for the next generation, the significance of having a low latency cannot be overstated. A low-latency connection is required for many applications, including virtual reality (VR) and augmented reality (AR), in order to prevent motion sickness in users and keep the user experience as immersive as possible. Telemedicine applications require almost instantaneous response times to ensure patient safety and successful procedures. This is analogous to the response times required for remote robotic surgery. Additionally, in the case of autonomous vehicles, there will be a need for instant communication between vehicles, infrastructure, and other users of the road. This is due to the fact that decisions will need to be made in a fraction of a second in order to reduce the likelihood of accidents and maximize the efficiency of traffic management.

The successful rollout and adoption of these cutting-edge applications will be contingent on the development of strategies for improving the dependability of 6G networks. One such method is known as "edge computing," and it entails the processing of data close to the data source at the edge of the network. By placing computational resources closer to users and devices, edge computing reduces latency, boosts network dependability, and strengthens data security. Additionally, edge computing helps in the offloading of processing workload from centralized data centers, which ultimately results in a network that is more dependable and efficient.

One more strategy for improving the dependability of 6G networks is to implement network slicing, a technology that makes it possible to divide a single network into multiple virtual networks. This capability makes it possible to create specialized network configurations that are catered to the specific requirements of various applications. This ensures the best resource allocation as well as the desired levels of latency, dependability, and quality of service for the network. Network slicing makes it possible to have control over the performance of the network at a fine-grained level, which paves the way for communication that is both extremely reliable and low in latency in 6G.

2.4 TECHNOLOGICAL DIRECTIONS FOR 6G

The 6G generation scheme is a combination of novel technologies and enhanced 5G architecture. This section discusses the new technologies which are disrupting the 6G. New technological developments, like blockchain, can improve the safety and proficiency of traditional spectrum (Tang, Zhang, & Li, 2023). MIMO and OAM can also improve spectrum efficiency (Li, Ge, & Zhang, 2022). Quantum communication (QC) and quantum machine learning (ML) can maintain the computation proficiency to maintain robust protection (Biamonte et al., 2023).

2.4.1 Artificial intelligence

Of the 6G technologies anticipated, AI is the most important and cutting edge. In contrast to the absence of AI in 4G systems, emerging 5G will allow for a restricted application. However, 6G will offer complete cooperation.

Machine learning development will lead to effective intelligent networks in 6G, enabling real-time transmission. AI is able to determine the most effective strategy to carry out challenging objective activities. AI will also decrease processing and communication latency, pick networks more effectively, and increase handover efficiency.

2.4.2 Blockchain technology

Another essential technology for managing 6G's voluminous traffic is blockchain. A database is spread across several computing units in a distributed ledger system called a blockchain. A duplicate ledger is kept by each processing unit. Since peer-to-peer networks manage it, a central server is not necessary. Blocks that are connected to one another and encrypted are used to compile and organize the blockchain data. With enhanced security, scalability, and dependability, it serves as a good substitute for the mMTC. As a result, it will offer a variety of services, including data traceability, IoT communication self-regulation, and 6G mMTC reliability.

2.4.3 Quantum communications

Unsupervised reinforcement learning is a powerful method for achieving the 6G visionary objectives. Contrarily, the huge amounts of data that will be created by 6G cannot be labeled using the supervised learning method. Complex networks can be created using unsupervised learning because labeling is not necessary. Additionally, the prospect of network operation in a truly autonomous manner is created by the combination of reinforcement learning and unsupervised learning.

2.4.4 Unmanned aerial vehicles

UAVs, sometimes known as drone BSs, are anticipated to play a significant role in the forthcoming 6G network (Alzenad et al., 2023). In a variety of circumstances, UAV technology will be employed to achieve effective wireless connectivity (Kakar et al., 2022). To do this, the BS troops will be sent out in UAVs. A UAV has benefits over a stationary BS, including simple placing and stringent line-of-sight requirements (Hayat et al., 2023). The deployment of a traditional terrestrial infrastructure-based network in the event of an emergency, such as a natural disaster, is not economically practical, and it is occasionally impossible to offer any service in such hazardous situations (Motlagh et al., 2023). UAVs, on the other hand, are able to handle such circumstances with ease because they satisfy the vital 6G specifications set forth by mMTC, eMBB, and URLLC (Bor-Yaliniz & Yanikomeroglu, 2022). UAVs can also be used for a wide range of other tasks, including surveillance, pollution monitoring, accident monitoring, network connectivity improvement, and fire detection.

2.4.5 3D networking

As a result of the 6G network's intended integration of several networks, users will be able to interact in vertical extensions. Additionally, the 3D BSs will be realized using UAVs. When compared to current 2D networks, the introduction of new altitude dimensions will drastically change 3D connection.

2.4.6 THz communications

One of the main objectives of every cellular generation is spectral efficiency, which can be further enhanced by expanding bandwidth. Spectral efficiency in 6G can be achieved by THz communication technology. The RF band can no longer support the growing 6G needs. The THz band is a good option to overcome this flaw in 6G since it is anticipated to be the future spectrum for data-hungry applications. THz waves have frequencies between 0.1 and 10 THz and wavelengths between 0.03 to 3 mm. ITU-R regulations state that the band between 275 GHz and 3 THz is appropriate for cellular communications because it is not globally designated for any other application and can deliver high data rates. This suggested section of the THz spectrum can increase overall 6G cellular communication capacity when coupled with the mmWave band between 30 and 300 GHz. The capacity of the entire band is increased by at least 11.11 times as a result of this addition.

2.4.7 Big data analytics

Enormous data analytics is a technique that is used to look into various big data collections. To ensure seamless data management, this technique makes information such as consumer preferences, occult patterns, and unexplained correlations visible. A variety of sources, including social networks, movies, sensors, and photographs, are used to gather the big data. This technology is said to be able to handle the huge amounts of data in 6G networks.

2.4.8 Holographic beamforming

An antenna array is used in the signal processing method known as beamforming to transmit radio signals in a directed manner. It is a collection of sophisticated antennas that provide many advantages, including increased network efficiency and little interference. Holographic beamforming (HBF), a cutting-edge method of achieving beamforming, differs significantly from MIMO systems in that it makes use of software-defined antennas. To establish successful communication in multiantenna communication network units, it is intended to be an efficient and effective 6G technology.

2.5 THE ONGOING 6G PROJECTS

2.5.1 Hexa-x

Ericsson unveiled the Hexa-x concept in 2021. In this partnership, numerous universities and research institutions work together to commercialize cutting-edge discoveries. The Hexa-x project aims to lay the groundwork for the 6G networks. It also aims to direct future generations' participation in international research and innovation (R&I). Enhancing the technologies necessary to bring 6G networks to Europe is the goal of this project. With the help of creative solutions that will be offered, six challenges—connected intelligence, a network of networks, sustainability, global service coverage, trustworthiness, and extreme experience—will be tackled. To focus on these challenges, Hexa-x will develop a number of axes. Human-device communication must take advantage of cutting-edge technology like AI and ML to increase connection quality. For the global digital ecosystem to exist, one network of networks must be established. This network should be knowledgeable, adaptable, and diverse. For a network to be sustainable, resources must be exploited effectively. Solutions that are workable and inexpensive must be developed in order for the 6G network to provide complete global coverage. In order to maintain high security, the next generation must assure data privacy, communication integrity, secrecy, and operational resilience. More technologies, such as network architecture, AI-driven air interface, THz radio access, and network virtualization, will be developed in order to enhance the performance of 6G. The goal of the project is to enhance communication between the physical, digital, and human worlds by concentrating on these cutting-edge communication technologies.

2.5.2 RISE 6G

RISE 6G (reconfigurable intelligent sustainable environments for 6G wireless networks) is one of the significant programs that will start in 2021. The project makes use of reconfigurable intelligent surfaces (RIS) technology. RIS will develop into one of the most potent developing technologies in the future. The dynamic aspect of controlling radio wave propagation is the subject of RIS. It makes it possible to see the wireless environment as a service. RISE 6G seeks to improve 6G capabilities for a flexible, intelligent, and sustainable wireless environment by leveraging RIS. The initiative will deal with four RIS-related problems (Strinati et al., 2021). The actual RIS-assisted signal propagation will be modeled first. Second, the new network architecture will enable the unification of numerous RISs. Third, a number of use cases will be created to support QoS. These use cases include massive capacity in a dynamic wireless programmable environment, green communication, reduced power consumption, and accurate localization. Fourth, an innovation prototype benchmark based on two complementary proceedings will be proposed. The project merges its technological vision into the application in industry and contributes to standardization.

2.5.3 New 6G

The NEW-6G project will be centered on the nano-world. The project connects "microelectronic with telecom," "network with equipment," and "software with hardware." In essence, the project will develop novel techniques and technologies to improve network performance, such as (Castro, 2022)

- Network planning and improvement;
- Data movement and protocols;
- Infrastructure and data security;
- Low energy consumption, integrated circuits, and high-performance radio frequencies;
- Environmentally friendly, committed, and efficient semiconductor technology; and
- NEW-6G will offer fresh applications for nanoelectronics technology.

We will look at nanoelectronics technology to generate fresh inquiries for both academia and industry.

2.5.4 Next G alliance

At the end of 2020, ATIS (Alliance for Telecommunications Industry Solutions) unveiled the Next G Alliance in the United States. ATIS aims to advance 6G leadership in North America by implementing the basics of 6G (Next G Alliance).

It concentrates on the technological commercialization process, which covers R&D, production, standardization, and market development. Future standards may be significantly influenced by member organizations' influence on important mobile communication players. The commercial trends and standards will be carefully considered by the Next G Alliance. We want to initiate a worldwide dialogue on standards and how industry and government can work together.

Mobile technologies are essential to the growth of many important businesses. The United States is becoming increasingly and more dependent on a variety of businesses as mobile technology advances, including aerospace, agribusiness, defense, education, healthcare, manufacturing, media, and transportation. North America must maintain its position as the global leader in mobile technology in these critical sectors.

REFERENCES

Agiwal, M., Roy, A., & Saxena, N. (2016). Next Generation 5G Wireless Networks: A Comprehensive Survey. *IEEE Communications Surveys & Tutorials*, 18(3), 1617–1655.

Akhtar, M. W., Hassan, S. A., Ghaffar, R., Jung, H., Garg, S., & Hossain, M. S. (2020). The Shift to 6G Communication: Vision and Requirements. *Human-Centric Computing and Information Sciences*, 10(1). 10.1186/s13673-020-00258-2

Alzenad, M., El-Keyi, A., Lagum, F., & Yanikomeroglu, H. (2023). 3-D Placement of an Unmanned Aerial Vehicle Base Station for Maximum Coverage of Users with Minimum Power. *IEEE Transactions on Wireless Communications*, 17(4), 2238–2251.

Biamonte, J., Wittek, P., Pancotti, N., Rebentrost, P., Wiebe, N., & Lloyd, S. (2023). Quantum Machine Learning. *Nature*, 549(7671), 195–202.

Boccardi, F., Heath, R. W., Lozano, A., Marzetta, T. L., & Popovski, P. (2021). Six Key Challenges for Defining 6G Wireless. *IEEE Communications Magazine*, 59(3), 82–88.

Bor-Yaliniz, I., & Yanikomeroglu, H. (2022). The New Frontier in RAN Heterogeneity: Multi-Tier Drone-Cells. *IEEE Communications Magazine*, 54(11), 48–55.

Castro, C. 6G Gains Momentum with Initiatives Launched across the World. *6GWorld*. 2022. Available online: https://www.6gworld.com/exclusives/6g-gains-momentum-with-initiatives-launched-across-the-world/

Ericsson. (2021). Ericsson Mobility Report: November 2021. Retrieved from https://www.ericsson.com/4ad539/assets/local/mobility-report/documents/2021/november-2021-ericsson-mobility-report.pdf

Fortune Business Insights. (2019). Internet of Things (IoT) Market Size, Share and Industry Analysis by Platform (Device Management, Application Management, Network Management), by Software & Services (Software Solution, Services), by End-Use Industry (BFSI, Retail, Governments, Healthcare, Others) and Regional Forecast, 2019–2026. Retrieved from https://www.fortunebusinessinsights.com/industry-reports/internet-of-things-iot-market-100307

GSMA Intelligence. (2021). The Mobile Economy 2021. Retrieved from https://data.gsmaintelligence.com/api-web/v2/research-file-download?id=34778545&file=MobileEconomy2021.pdf

Hayat, S., Yanmaz, E., & Muzaffar, R. (2023). Survey on Unmanned Aerial Vehicle Networks for Civil Applications: A Communications Viewpoint. *IEEE Communications Surveys & Tutorials*, 18(4), 2624–2661.

Khalid, A., Saeed, A., & Raza, S. (2021). A Survey on Terahertz Spectrum for 6G Wireless Communication. *Journal of Network and Computer Applications*, 181, 103053.

Kakar, P., Jha, R. K., & Jain, S. (2022). A Comprehensive Survey on Mobile Edge Computing: Architecture, Functionality, Use-Cases, Motivation, Issues and Research Challenges. *Journal of Network and Computer Applications*, 134, 55–69.

Li, Y., Ge, X., & Zhang, Y. (2022). 6G Wireless Communications: Vision and Potential Techniques. *IEEE Network*, 32(2), 70–75.

Motlagh, N. H., Taleb, T., & Arouk, O. (2023). Low-Altitude Unmanned Aerial Vehicles-Based Internet of Things Services: Comprehensive Survey and Future Perspectives. *IEEE Internet of Things Journal*, 3(6), 899–922.

Next G Alliance FAQ. ATIS. Available online: https://nextgalliance.org/about/

Nawaz, S. J., Sharma, S. K., Wyne, S., & Patwary, M. N. (2021). Quantum Machine Learning for 6G Communication Networks: State-of-the-Art and Vision for the Future. *IEEE Access*, 9, 3171–3190.

Pouttu, A., Burkhardt, F., Patachia, C., Mendes, L., Brazil, G.R., Pirttikangas, S., Jou, E., Kuvaja, P., Finland, F.T., & Heikkilä, M. (2020). 6G White Paper on Validation and Trials for Verticals Towards 2030's. *6G Research Visions*, 4.

Qualcomm. (2021). The 5G Economy: How 5G Technology Impact Will Contribute Trillions to the Global Economy. Retrieved from https://www.qualcomm.com/media/documents/files/the-5g-economy.pdf

Saad, W., Bennis, M., Chen, X., Hong, C. S., Kiani, A., Larsson, E., ... & Guvenc, I. (2021). A Vision of 6G Wireless Systems: Applications, Trends, Technologies, and Open Research Problems. *IEEE Network*, 35(3), 134–142.

Strinati, E., Alexandropoulos, G. C., Wymeersch, H., Denis, B., Sciancalepore, V., D'Errico, R., Clemente, A., Phan-Huy, D.-T., De Carvalho, E., & Popovski, P. (2021). Reconfigurable, Intelligent, and Sustainable Wireless Environments for 6G Smart Connectivity. *IEEE Communications Magazine*, 59, 99–105.

6G White Paper on Validation and Trials for Verticals towards 2030's; 6G Research Visions. 2020. Available online: https://www.6gchannel.com/items/6g-white-paper-validation-trials/

Tang, Z., Zhang, J., & Li, K. (2023). Blockchain-Enabled Spectrum Sharing in Cognitive Radio Networks. *IEEE Transactions on Vehicular Technology*, 67(8), 7042–7054.

Vreman, N., & Maggio, M. (2019, April 15). Multilayer Distributed Control over 5G Networks: Challenges and Security Threats. In Proceedings of the Workshop on Fog Computing and the IoT, New York, NY, USA; pp. 31–35.

Chapter 3

Ultra-dense deployments in next-generation networks and metaverse

Ravi Sekhar Yarrabothu[1] and G. Ramana Murthy[2]
[1]Department of ECE, VFSTR, Vadlamudi, India
[2]Department of ECE, Alliance University, Bangalore, India

3.1 INTRODUCTION

Fifth Generation (5G) networks will provide enormous capacity, increased reliability, and low latency. However, there is still a long way to go before 5G networks can fully support applications like the Internet of Everything (IoE), collaborative robots (cobots), holographic telepresence (HT), and deep-sea and space travel. Ultra-high data rates, ultra-low latency, ultra-high dependability, ultra-high-speed processing, extremely high connection densities, and ubiquitous coverage – over land, in the air, in space, and underwater – are just a few of the limits. The sixth-generation (6G) mobile network is projected to utilize a wider frequency spectrum in terms of giga-hertz bandwidth, cost-effective technology, and a higher degree of reli-ability to improve coverage and reduce battery power usage. Several enabling technologies, including more modern waveform designs, effective multiple access methods, sophisticated channel coding techniques, massive MIMO antenna technologies, network slicing, and cloud edge computing, are used in 6G networks. Table 3.1 is a quick summary of the 6G criteria and use cases.

6G influences four significant future development aspects (Alwis et al., 2021). These features include global coverage, numerous spectra, innumer-able new applications and services which were never thought of before due to advances in AI and ML, and stronger security. The first aspect of global coverage the of communication network involves a centralized ground, air, space, and underwater network by integrating the terrestrial and non-terrestrial (NT) networks (Ray et al., 2021), as shown in Figure 3.1.

The second aspect is an institution of supplementary radio frequency bands, which includes millimeter-wave (mm wave), optical communica-tions, and terahertz (THz) frequencies, to improve the capacity and efficiency of the communication networks. The third aspect denotes the completely new applications and services using artificial intelligence (AI), machine learning (ML), and big data analytics, which leads to the

DOI: 10.1201/9781003369028-3

Table 3.1 Overview of 6G requirements and use cases

	6G requirement	*Use cases*
Data rates	More than 1 TB of data rates	3D holographic, AR/VR, Robotics Arm
Coverage	3D coverage scenarios (10,000 m in Sky), 200 NM (sea)	Terrestrial, aerial, space and sea domain, massive-scale IoT network
Battery power consumption	50 times improvements compared to 5G, nearly (1 Tb/J)	Wearable user devices, zero energy devices
Latency (end-to-end)	100 microseconds to 1 ms	Healthcare networks, AR/VR, unmanned ariel vehicle (UAV), robotics arm
Communication reliability	~ 99.9999%	Healthcare networks, AR/VR, unmanned ariel vehicle (UAV), robotics arm
Massive connectivity and sensing	10 million devices/km^2	Wearable user devices, AR/VR, IoE
Frequency spectrum	Up to 1 THz	mmWave, Sub-6 GHz, exploration of THz bands (above 300 GHz), high-definition imaging and frequency non-RF (e.g., optical, VLC), spectroscopy, localization
Mobility and speed supportive	Up to 1,000 km/hr	Terrestrial, space, sea, aerial, and airline

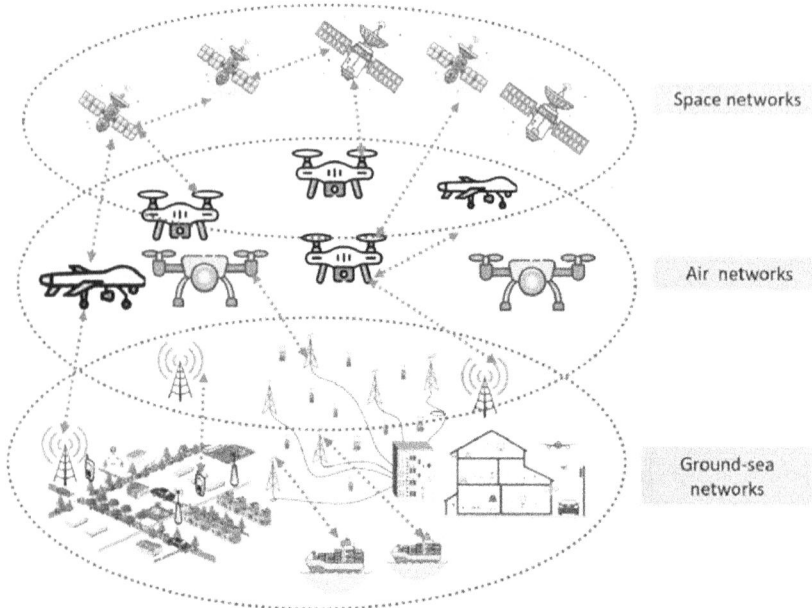

Figure 3.1 The integrated ground-sea-airspace 6G expected network.

development of wide-ranging bandwidth utilization algorithms and MIMO antenna designs that are required for 6G technical specifications. The fourth aspect is to provide robust security to the data that is exchanged, which leads to the improvement the network confidentiality.

The IoE presents a very challenging issue of data security and privacy issues as well as reliability issues. In 6G, it is anticipated that the end-to-end latency may be in the order of a few microseconds for the enhanced ultra-reliable and low-latency communication (URLLC) services. In 6G communication networks, extreme applications such as unmanned aerial vehicle (UAV) communication, HT, extended reality (XR), smart grid 2.0, Industry 5.0, and space and deep-sea connectivity will become normal applications. However, the requirements of these extreme applications are beyond the network capabilities promised by 5G. 6G networks are likely to make a disruptive revolution to mobile networks by creating extreme network capabilities to accommodate the necessities of the upcoming data-centric society. By 2030, society may have wireless connectivity as the communication and information pillar, letting ubiquitous communication happens anywhere and anytime. Sustainability is of extreme importance to all parts of society, which needs to keep its efforts towards the United Nations Sustainable Development Goals. Some of the existing wireless networks play a crucial role in achieving these development goals, and there is a strong possibility of further acceleration in empowering better utilization of the resources, supportive of new ways of living and building them as an instrument for positive change.

6G technology creates an opportunity to move in a cyber-physical continuum, among the associated physical world of senses, actions, and experiences and its programmable digital representation as shown in Figure 3.2. The 6G networks provide smartness, unlimited connectivity, and fully synchronized physical and digital worlds. The physical world would be sending the data from the innumerable number of sensors that are embedded to update the digital representation in real time. The actuators in the real world would be executing the commands from intelligent agents of the digital world. The cyber-physical continuum offers a close connection to realizm, where the digital twins are created which are the digital representation of physical objects, letting them smoothly coexist as merged reality and enhance the real world.

For thorough event planning, cyber-physical platforms can provide interactive 4D maps of the entire city that are accurate in place and time and can be accessed and altered at the same time by large numbers of people and intelligent devices. To achieve higher levels of resource efficiency, superior control, and more flexibility, such cyber-physical service platforms can issue orders to large-scale steerable systems, such as public transportation systems, waste management systems, or utility management systems. To control and monitor the heavy traffic in the next smart cities using on-the-ground autonomous driving vehicles and aerial autonomous drones, interactive real-time 4D maps are essential. The measurements and real-time data

Cyber-physical continuum

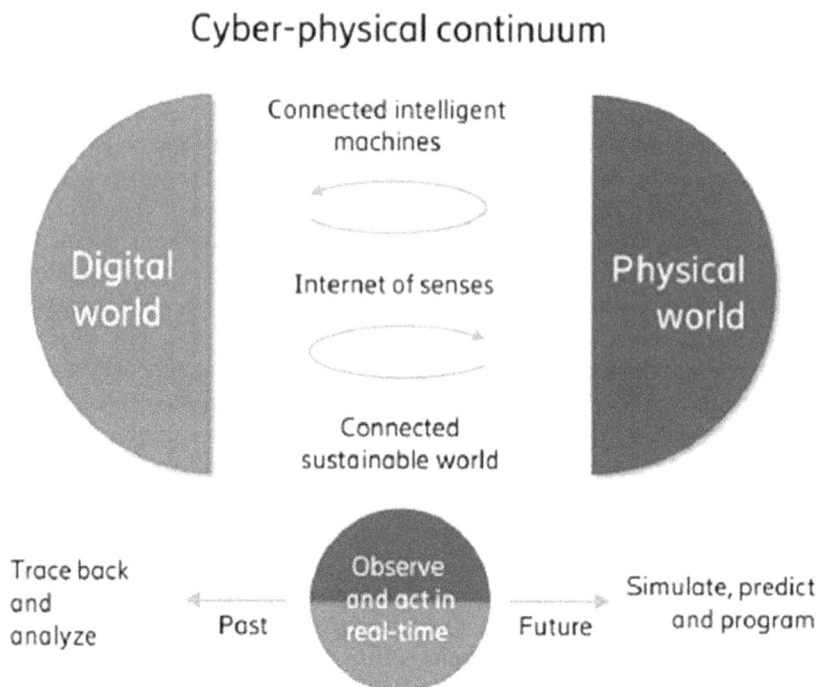

Figure 3.2 Cyber-physical continuum.

are gathered from the network of onboard sensors in cars via the base stations (BSs) to create a safer, more effective, and well-guided transportation system.

Immersive communication is going to bring the experience of full telepresence, by eliminating the remoteness hindrance for interactive communication. XR technology like human sensory feedback necessitates extreme data rates and volume, spatial mapping from accurate positioning and sensing, and ultra-low latency end-to-end with edge cloud processing. In the future, communication needs to support the complete merger of physical senses and digital. Personal immersive devices, which can provide precise body interaction create experiences and actions remotely, guaranteeing an immersive perception, to cater to the needs of better communication between people with emotions, whose prominence is pretty much clear during the COVID-19 pandemic time.

3.2 ULTRA-DENSE DEPLOYMENTS IN 6G

The upcoming 6G wireless networks should offer truly extreme performance, including terabits/sec data speeds and microsecond latency performance, when necessary, extreme system capacity for providing the services to many users at once, and complete global coverage of the wireless access. The use of more modern packet front haul and wireless transport technologies, free-space optics, and more integrated access and backhaul are the keys to providing dense deployments with enormous system capacity in a way that is both efficient and cost-effective. To give continuous coverage everywhere for 6G wireless connectivity, expanding the standard terrestrial access to incorporate NT access elements such as drones, high-altitude platforms, and/or low-Earth orbit (LEO) satellites. Trillions of intelligent gadgets must be supported by 6G networks, which also need to offer dependable connections that are always available.

The IoE devices and BS densification are major components of network densification in 6G wireless networks. The unique characteristics of network densification make the currently used conventional network solutions inefficient and would also bring forth new difficulties. In this article, new issues and prospective important technologies are introduced after the characteristics of network densification in terms of both BS and ED densification are first investigated.

The enormous growth in traffic demand of mobile users has imposed critical requirements for 6G wireless communications, such as ultra-high data speeds and ultra-low latency while offering ultra-wide radio coverage. This has led to the identification of ultra-dense networks (UDNs) as a potentially efficient method to meet these 6G standards. Massive and dense deployments of small BSs (e.g., over 1,000 BSs/km^2) are projected to be realized in UDN for increasing data throughput and coverage. Significant study attention has been paid to characterizing the effects of BS densification on the performance of wireless networks (Gupta et al., 2021). To meet the future demands of information and communication technology (ICT), the sixth generation (6G) plays a vital role. The 6G system is anticipated to provide ten times more capacity and ten times lower latency than the 5G system while supporting ten times more connectivity, as shown in Figure 3.3. The end-device (ED) densification caused by the exponential rise in machine-type devices (MTDs) in 6G will make it challenging for the 5G system to handle. In contrast to the traditional network, a variety of EDs, such as human-type devices (HTDs) and machine-type devices (MTDs), such as Internet of Things (IoT) sensors, automobiles, and UAVs, would repeatedly request service from the network.

Densification of BSs in 6G wireless communications brings forth brand-new difficult issues. The distance between the BSs decreases as the BS density rises (of the order of a few meters). As a result, there is a decreased distance between the interfering BSs and the target user, which leads to high interference power and low coverage probability. Additionally, certain BSs will have no related ED as the density of the BSs rises, causing wasteful network power usage. Furthermore, the dense network environment is anticipated to have multiple tiers of BSs to support different coverage ranges and functionalities as well as for seamless integration into the

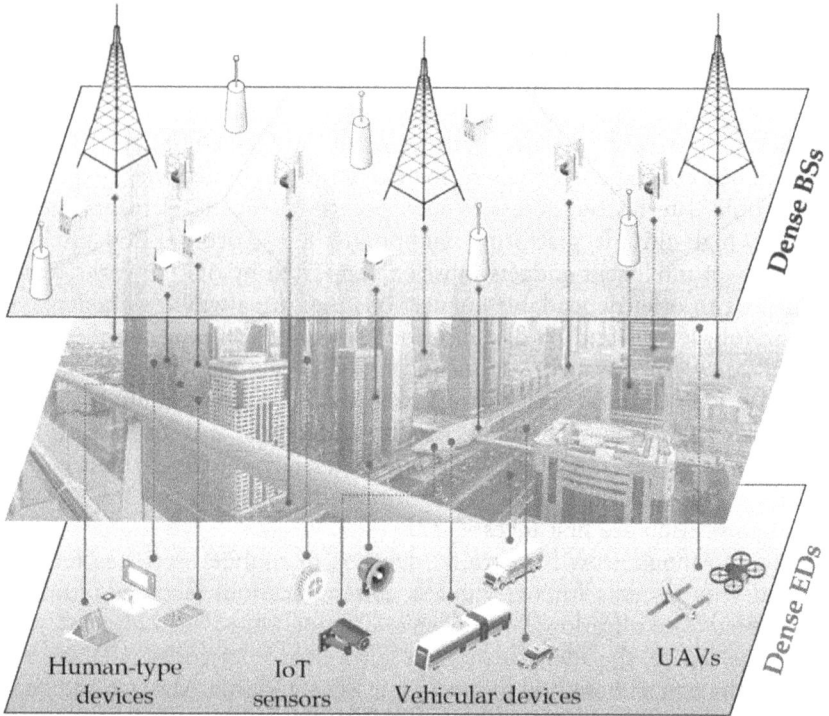

Figure 3.3 Network densification with various types of BSs and EDs.

current cellular networks, similar to the conventional heterogeneous networks (HetNets), which are networks consisting of different types of access nodes. However, compared to a traditional multi-tier network that does not take into account UDN aspects, such as the non-negligible impact of BSs having varying antenna heights, the performance of a multi-tier dense network will be different. The best architecture for the multi-tier dense network, therefore, presents a new challenge.

Additionally, ED densification creates several brand-new difficult issues. MTD traits are entirely distinct from those of traditional HTD traits. Traffic that might be regular or erratic is one of the new elements in MTD communications. Despite the great density of MTDs, these traffic patterns only permit a limited number of devices to be active at once. As a result, it becomes crucial to deal with precisely detecting the active devices. Additionally, the spatial correlations in interference and main link channels are no longer insignificant when the space between EDs gets significantly closer. The key characteristics of network densification in terms of BS and ED densification are outlined, and several topics and technologies are introduced, as illustrated in Figure 3.4, to meet these opportunities and difficulties of network densification. The opportunities provided by network densification in 6G wireless networks are then discussed.

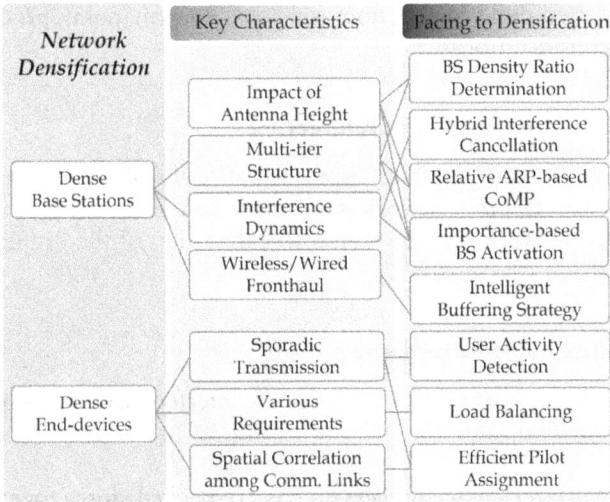

Figure 3.4 Key characteristics and challenges of network densification.

3.3 KEY CHARACTERISTICS OF NETWORK DENSIFICATION

The next 6G wireless network differs from the current network in terms of features due to the densification of both BSs and EDs. The key network properties of the 6G wireless network derived from BS densification and ED densification, respectively, are covered in the sections that follow.

3.3.1 BS densification's key network properties

3.3.1.1 Significant impact of antenna height

The nature of line-of-sight (LoS) and non-line-of-sight (NLoS) propagation is affected by several critical parameters, one of which is the height difference between the ED antenna and its associated BS antenna. For instance, the propagation connection may be an LoS or NLoS link based on the height difference between the BS and the ED and the height of the blockage situated between them. The statistical model for calculating the geometrical likelihood of a signal between a transmitter and a receiver was developed. The trend of the LoS probability can be closely compared to a sigmoid curve. Additionally, if an LoS propagation is present, as opposed to if an NLoS link is present, the main and interference link channels can be modeled with Rician or generic Nakagami-m fading with decreased path loss exponents, which are known to boost the received signal strength (i.e., Rayleigh fading only). It has been demonstrated that there is an ideal BS antenna height for a particular BS density that would maximize the network performance due to the non-negligible impact of the antenna height variation. The single-tier network with single-antenna BSs and single-antenna EDs dispersed spatially randomly by separate two-dimensional homogeneous Poisson point processes was taken into consideration.

3.3.1.2 Multi-layer network architecture

A multi-layer network topology is a natural result of the BS densification process. However, if the network is not intricately built, the performance of the multi-tier network may be even lower than that of the ordinary single-tier network.

3.3.1.3 Dynamics of Interference

The following can be used to summarize the interference aspects brought on by BS densification:

- *Stronger interference power:* As BSs are packed closer together, more BSs can interfere with an ED in the network, but there are also fewer BSs between the BSs that can interfere and the ED itself. This also raises the possibility of a network having an LoS-interfering link. In contrast to traditional sparse cellular networks, the interference from interfering nodes is hence stronger in a dense network. Additionally, in a multi-tier network, BSs from several tiers are the source of the inter-cell interference (i.e., BSs with different antenna heights and transmission powers). In other words, due to BS densification, the received interference varies depending on the tier of the interfering BS.
- *Spatially correlated interference and main link channels:* As the distance between BSs decreases, the spatial channel between the serving BS and its interfering BSs becomes comparable, causing both the interfering link channels and the main link channels to exhibit significant spatial correlation.
- *Constant change in the interfering node set:* Some BSs may not have any linked EDs due to the higher number of different BSs and EDs with erratic traffic patterns. (i.e., no data to transmit). In contrast to traditional networks, the BS changeover might occur more frequently. The set of interfering BSs can vary more often over time due to the network's increased dynamism.

3.3.1.4 Wireless and wired fronthaul coexistence

The link quality and throughput between the BS and the ED may be improved by the BS densification. However, when network traffic grows, it becomes important to use a high-capacity, low-latency backhaul to connect the increasing number of BSs to the backbone network. All macro-BSs are directly connected to the wired backbone network of the conventional network. Such a wired backhaul connection arrangement is no longer effective or viable as the BS density rises. Instead, the use of macro-BSs, often referred to as fronthaul, to connect BSs, particularly small BSs, with the backbone network has been encouraged (Yaacoub & Alouini, 2020). It might not be possible to manage all small BSs with wired fronthaul due to BS densification, though. For instance, using the wired fronthaul is impossible when UAV BSs are taken into account. Therefore, the dense 6G network requires wireless fronthaul (Jiang et al., 2022).

Due to channel fading and fluctuation, the wireless fronthaul has a lower capacity than the cable fronthaul. In particular, the channel conditions have a significant impact on the wireless fronthaul link's (and wireless access link's) capacity. Due to the capacity differential between the wireless fronthaul and the access link, some information may be lost at the macro-BS if the channel condition at the wireless fronthaul is poorer than that of the access link (Polese et al., 2020). The throughput of the ED will be constrained by the fronthaul capacity, for instance, if the attainable rate of the wireless fronthaul is lower than that of the access link. By using a buffer at the small BSs, a buffering technique can be used to reduce the effects of the constrained fronthaul capacity and solve the information loss problem. Small BSs can store the data or information with the buffer rather than dropping it.

3.3.2 Significant network properties from ED densification

3.3.2.1 Intermittent transmission of devices

In contrast to traditional HTDs, MTD traffic is typically irregular. To put it another way, ED densification has resulted in a huge number of EDs, including both HTDs and MTDs, yet only a small portion of the MTDs is active at any given time. Due to this property, it is challenging for the BSs to identify the transmitting EDs. As a result, to maximize network performance, the receiver must first locate the actual active ED set using an active ED detection methodology before properly decoding any messages sent by that set using the appropriate decoding method.

3.3.2.2 Coexistence of various sorts of traffic

Along with the traditional HTDs, several kinds of MTDs also appear in the network because of ED densification (Mahmood et al., 2020). There can be variations in the requirements for certain EDs. For instance, while certain MTDs, such as basic IoT sensors, require best-effort connectivity and recent data with the low information age, conventional HTDs and some types of MTDs, such as automobiles, require high data rates and highly dependable real-time connectivity (AoI). To maximize network performance, it is important to take into account the coexistence of different traffic kinds with diverse service requirements. This calls for the best resource allocation and scheduling across EDs.

3.3.2.3 Environment with a spatial correlation between EDs

Additionally, spatial correlation is produced by ED densification. It takes into account both the correlation of regional ED information and the correlation of channels with BSs. The channel components are increasingly similar, or spatially correlated, the closer two ED locations are to one another. Thus, ED densification will result in a greater rise in the spatial correlation between wireless channels. The likelihood of the ED associating with the BS that serves other EDs around the ED is high, however, as the distance between EDs reduces with ED densification. The performance of the network can be further enhanced by taking advantage of the spatial correlation between nearby EDs. For instance, the estimated value of the next ED channel may be substituted for or inferred from the estimated value of this ED's channel.

3.4 FACING THE DENSIFICATION OF NETWORKS

The main difficulties with densifying BSs and EDs are covered in this section.

3.4.1 Facing BS densification

When there are different types of BSs, which have also been researched in the HetNet environment, there may be a few difficulties arising from the densification of BSs. A multi-tier dense network, however, introduces unique difficulties when switching from a traditional multi-tier network. The following new issues that have arisen specifically as a result of BS densification, as opposed to HetNet, will be the focus of this section: (a) coordinated multi-point (CoMP) design (Mukherjee et al., 2021), (b) hybrid interference management (Zhan & Dong, 2021), and (c) BS activation control (Park et al., 2021).

3.4.2 Facing ED densification

3.4.2.1 Issues with pilot contamination and user activity monitoring

One of the key difficulties in massive MIMO, also known as multi-user multiple-input multiple-output, is the precise channel estimation. However, due to ED densification, there could not even be enough orthogonal pilot sequences to distinguish the uplink channel estimation from various EDs as the number of EDs increases significantly. This makes the pilot contamination problem with ED densification inevitable.

However, one of the key differences between machine-type communication (MTC) traffic and normal human-type communication (HTC) is that it is intermittent. To put it another way, even though the number of ED rises due to ED densification, the actual number of EDs that are actively interacting at any given time is substantially lower. Therefore, the limited orthogonal resources can be distributed to the real transmitting EDs more effectively by accurately identifying or forecasting user behavior.

3.4.2.2 Fronthaul capacity dependent on available BS association

It is demonstrated in Section 3.3.1.4 that BS densification makes wireless front haul between macro and small BSs inevitable. To further improve network performance and mitigate the consequences of the wireless fronthauls' low capacity, a fronthaul capacity-based BS association mechanism is considered. The strongest ARP or the shortest link distance are the major objectives of traditional BS association tactics. However, some BSs may be overloaded when used with traditional BS association systems because of the constrained fronthaul capacity. Because of the imbalanced load, BSs will bottleneck and the performance of the entire network will suffer. To solve this problem, serving BS is chosen by taking into account the BSs' available fronthaul capacity. The ED may associate with the BS that is anticipated to provide higher and more reliable communication despite the limited fronthaul capacity by taking into account not only the connection distance and ARP but also the available fronthaul capacity of the BS at the present instant.

3.5 POSSIBILITIES PRESENTED BY NETWORK DENSIFICATION

In this section, the focus is on the advantages that network densification has to offer.

3.5.1 Making BSs functional for different services

A variety of BS types will be deployed in the network because of BS densification, as detailed in Section 3.3.1.2. Each BS may have a variety of roles, including BSs with processing and caching capabilities and BSs focused on aerial devices, to fully exploit these densely dispersed BSs.

3.5.1.1 BSs with computing abilities

Some level of computation can be processed at the network edge, such as BSs, in the multi-tier dense network environment with edge computing, as opposed to needing computational support from the network core, such as a cloud server. The degree of computing and communication resources, such as CPU frequency and transmission power, would vary between each layer of BS. Higher-tier BSs, such as macro-BSs and first-tier BSs, typically have more transmit power and compute power than BSs at lower-tier BSs (e.g., small BSs, 2nd-tier BSs). As opposed to higher-tier BSs, lower-tier BSs could offer superior communication performance but worse computation performance. Therefore, choosing a BS that takes into account both communication and computation performance will be problematic, as opposed to one that solely takes into account communication performance. In addition to enhancing computing performance, creating an intelligent offloading strategy to take advantage of the multi-tier dense network topology will also enhance the ED's quality of services and experiences.

3.5.1.2 BSs with cache functionality

By using the caching strategy, BSs can store popular data like the most recent news and adverts, and EDs can instantly access required data from adjacent BSs, lowering latency. These days, the idea of service caching has also been proposed, which caches not only data but also service software that may run and calculate a particular application. A significant number of BSs will be able to cache and compute the data or service software as a result of BS densification. The hit probability will rise as a result, and more service algorithms will be accessible at the network's edge. Consequently, the network performance will improve with appropriate data and service cache techniques devised by the characteristics of the many BSs in a multi-tier dense network.

3.5.1.3 Aerial device-oriented BSs

The emergence of aerial users (such as UAVs) as well as typical ground users should be considered when ED densification is being done. A unique BS service scheme is needed to cater to both ground users and aerial users. In particular, an exclusive-service BS scheme has BSs for ground users and BSs for aerial users exclusively while an inclusive-service BS scheme makes BSs serve both ground users and aerial users simultaneously. To improve network performance from BS densification, the deployment of BSs utilizing each scheme must be properly planned.

3.5.2 Utilizing geographical correlation between communication lines to assign pilots

Traditionally, an orthogonal pilot signal sent from the EDs was used to accomplish channel estimation at the BS. However, as a result of ED densification, there are more ED connected to the BS than there are accessible orthogonal pilot sequences. An innovative channel estimation scheme derived from network densification can be used to address the pilot shortage issue. Channel estimate of a portion of EDs connected to the BS will be carried out using the traditional orthogonal pilot signals by utilizing the spatial correlation characteristic of EDs, which was explained in Section 3.3.2.3. Following that, as the surrounding EDs are spatially linked, the channel information of those EDs may be inferred based on the estimation result. The pilot shortage issue can be solved while performing densely deployed ED channel estimation by making use of the spatial correlation feature.

3.5.3 Developing an intellectual buffer method to reduce the fronthaul limit

The backhaul capacity in a traditional network with cable backhaul was thought to be consistently higher than the wireless access lines. The fronthaul capacity, however, may be less than the access link due to the wireless front hauls constrained capacity in the multi-tier dense network. In particular, the throughput of the tiny BS ED will be constrained by the fronthaul capacity if the attainable rate of the wireless fronthaul is lower than that of the access connection. The buffer technique in a multi-tier dense network is one of the potential remedies to lessen the effects of the wireless front hauls restricted capacity. When the fronthaul capacity is greater than the access link capacity, the small BSs can hold the data instead of dropping it by using a buffer. Additionally, the tiny BS can send its stored data in its place when the fronthaul capacity is less than the capacity of the access link. In the multi-tier dense network, an advanced technique like a reinforcement learning algorithm can be used to control the buffer strategy effectively.

3.6 METAVERSE

The advent of the IoE and the ability to connect people, devices, and the cloud from any location is made feasible by 6G-enabled networks. Smart service applications for the next-generation wireless networks are transforming our way of life and raising our standard of living. The goal of the metaverse, the hottest new category of next-generation internet apps, is to link billions of people and develop a shared environment where virtual and real worlds coexist (Chang et al., 2022). The idea of the metaverse has gained momentum as "the successor to the mobile Internet." The whole concept of an immersive, embodied, and interoperable metaverse has not yet been fully realized, even if there are lighter versions of the metaverse available today (Xu et al., 2023).

The metaverse is unlikely to replace the internet without addressing implementation concerns from the communication, networking densification, and computation viewpoints, particularly given its current accessibility to billions of people. In this survey, the main focus is on the edge-enabled metaverse to realize the metaverse's purpose and in section 3.5, the primary enabler of the ultra-densification of networks is discussed. A description of the architecture and a quick overview of the metaverse is presented first, along with a list of recent developments. The focus shifts to communication and networking issues to enable ubiquitous, seamless, and embodied access to the metaverse, and it examines cutting-edge solutions and concepts that make use of next-generation communication systems to enable users to engage with and immerse themselves as embodied avatars in the metaverse. Given the high computation costs necessary, for example, to render 3D virtual worlds and run data-hungry AI-driven avatars, the computational challenges and cloud-edge-end computation framework-driven solutions to realize the metaverse on resource-constrained edge devices are also covered. The next section provides insight into how blockchain technology might facilitate the interoperable expansion of the metaverse, not only by facilitating the commercial distribution of digital user-generated content but also by managing physical edge resources in a decentralized, open, and unalterable manner. This chapter concludes with a discussion of the potential research routes for achieving the true goal of the edge-enabled metaverse.

3.6.1 Emergence of the metaverse

Neal Stephenson's 1992 science fiction book *Snow Crash* introduced the idea of the metaverse for the first time. The term "Metaverse" has returned to the lexicon after more than 20 years. Users will employ augmented reality (AR), virtual reality (VR), and the tactile internet to explore the virtual worlds within the metaverse, like how currently web pages are surfed with a mouse cursor. The metaverse is currently being developed by IT firms as "the successor to the mobile Internet". The metaverse eventually surpasses the internet in terms of transforming new service ecosystems in all spheres of human endeavor, including healthcare, education, entertainment, e-commerce (Jeong et al., 2022), and smart industries. Facebook changed its name to "Meta" in 2021 as it transformed from a "social media corporation" to a "metaverse company" to reaffirm its dedication to the growth of the metaverse.

The excitement surrounding the metaverse is primarily driven by two factors. First off, the COVID-19 pandemic has caused a paradigm shift in how people currently engage in work, enjoyment, and socializing (Bates et al., 2021). The metaverse has been positioned as a requirement in the foreseeable future as more people become acclimated to performing these traditionally physical activities in the virtual realm. Second, the metaverse is becoming a more real prospect thanks to newly developed technical enablers. Beyond 5G/6G (B5G/6G) communication systems, for instance, promise eMBB (enhanced mobile broadband) and URLLC (ultra reliable low latency communication) (Shen et al., 2021; Tang et al., 2022), both of which enable AR/VR and haptic technologies that allow users to be both visually and physically immersed in the virtual worlds.

The metaverse is viewed as the long-term goal and advanced stage of digital transformation (Yuan et al., 2022). The metaverse, an outstanding multi-dimensional and multisensory communication medium breaks the shackles of distance by allowing users in various physical locations to connect and immerse themselves in a common 3D virtual reality (Duan et al., 2021). The "lite" forms of the metaverse that exist today are primarily derived from massive multiplayer online (MMO) games. Cities throughout the world have started ambitious attempts to create their meta-cities (Zhou et al., 2022) in the metaverse outside of the game industry. However, there is still a long way from achieving the metaverse's full potential (Lim et al., 2022). The physical and virtual worlds are anticipated to be integrated by the metaverse, for example by using the digital twin (DT) to reproduce the physical environment virtually (Wu et al., 2021). The real-time and scalable implementation of the metaverse is hampered by the strict sensing, communication, and processing requirements. Finally, the development of the metaverse coincides with a tightening of privacy laws. A data-driven understanding of the metaverse raises new issues since new ways of using AR and VR to access the internet suggest that new modalities of data, such as eye-tracking data, can be gathered and used by corporations.

The edge devices, or those connected to the metaverse via radio access networks, may not be able to support interactive and resource-intensive applications and services, such as interacting with avatars and rendering 3D worlds, even though the metaverse is thought to be the internet's replacement. People must be able to access the metaverse from anywhere, much as the internet amuses billions of users every day, for it to succeed as the next-generation internet. Fortunately, it has become clear that upcoming communication systems will be developed to support the metaverse because AR/VR, the tactile internet, and hologram streaming are key driving applications of 6G. Furthermore, the timely shift away from the conventional focus on traditional communication metrics like data rates and toward the co-design of computation and communication systems suggests that the design of next-generation mobile edge networks will also make an effort to address the issue of bringing the metaverse to computationally limited mobile users. Last but not least, the paradigm shift from centralized big data to distributed or decentralized little data throughout the "Internet of Everything" suggests that the blockchain (Huynh-The et al., 2023) plays a crucial role in achieving the metaverse at mobile edge networks. In light of the aforementioned factors, this survey concentrates on talking about the edge-enabled metaverse. To resolve these concerns advancements in computer and networking, communication, and blockchain/distributed ledger technology (DLT) will be crucial.

3.6.1.1 Contributions and related works

Numerous surveys on similar subjects have lately been published (Lee et al., 2021; Huynh-The et al., 2023) due to the metaverse's popularity. A synthesis of these surveys offers a tutorial that motivates a description and introduction to the architecture of the metaverse and identifies major networking and communication issues that must be resolved before the immersive metaverse can be realized. Major communication and networking issues that must be resolved are discussed, along with ideal metaverse characteristics. The survey also outlines major computational challenges for resource-constrained users at the network edge to realize the ubiquitous metaverse, the contribution of blockchain technologies to the creation of the interoperable metaverse, and the research directions to set the stage for upcoming efforts to realize the metaverse at mobile edge networks.

3.6.2 Architecture and tools of the metaverse

The architecture, use cases, and development resources for the metaverse are outlined in this section.

3.6.2.1 Definition and architecture

The metaverse is an embodied version of the internet that combines seamlessly navigable user avatars with shared, immersive, and interoperable virtual ecosystems. As depicted in Figure 3.5, the metaverse's architecture is explored together with some key enabling technologies.

1. **Physical-virtual synchronization:** Every non-exclusive stakeholder in the real world has control over elements that have an impact on the virtual realms. Stakeholder behavior has effects in the virtual world that could have an impact on the real world as well. The main parties involved are:
   - *Users* can fully immerse themselves in virtual worlds as avatars through various devices, such as head mounted displays (HMDs) or AR goggles. In virtual communities, users can interact with real or virtual items by executing actions.
   - *The IoT and sensor networks* installed in the real world gather data from that environment. For example, the resulting insights are utilized to maintain DTs for physical entities in the virtual worlds. Sensing service providers (SSPs) that give live data feeds to virtual service providers (VSPs) to create and manage virtual worlds may separately own sensor networks. This is known as sensing as a service.
   - *VSPs* create and support the metaverse's virtual worlds. User-generated content (UGC), such as games, art, and social applications, is expected to enrich the metaverse, much like user-uploaded movies do now (such as those on YouTube).
   - *Physical service providers (PSPs)* run the metaverse engine's supporting physical infrastructure and also respond to metaverse-generated transaction requests. The delivery of tangible items

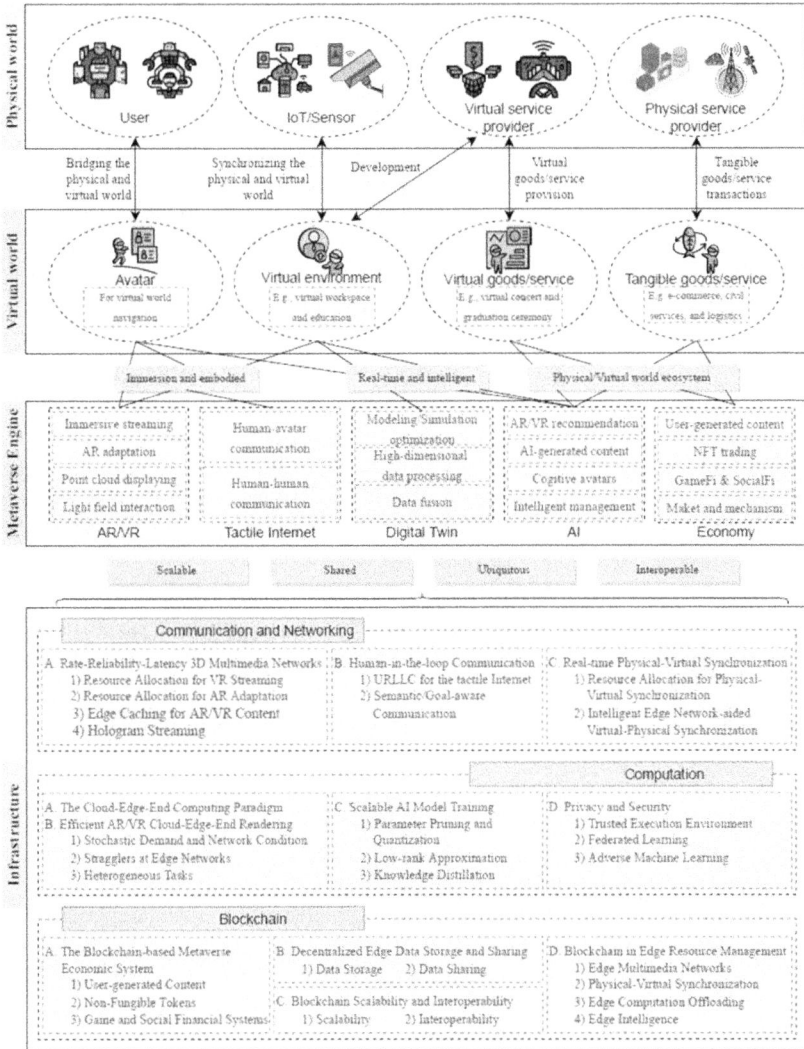

Figure 3.5 The physical and virtual worlds synchronization in real time through the metaverse architecture.

purchased in the metaverse also includes the use of communication and computing resources at edge networks or logistics services.

2. **The metaverse engine:** It receives inputs from entities and their activities in the real and virtual worlds, including data from stakeholder-controlled components that are developed, maintained, and improved.

- *AR/VR* allows users to experience the metaverse visually; haptics, such as haptic gloves, allow users to experience the metaverse in addition to sight and sound. This improves user interactions, such as when a handshake is sent across the globe, and also makes it possible to offer actual services in the metaverse, like remote surgery. These technologies are created by interoperability standards, such as virtual reality modeling language (VRML), which regulate the characteristics, physics, animation, and rendering of virtual assets and enable seamless user movement within the metaverse.

- *Tactile internet* technology allows metaverse users to send and receive haptic and kinesthetic information with a round-trip delay of about one millisecond. For instance, Meta's haptic glove project (https://about.fb.com/news/2021/11/reality-labs-haptic-gloves-research/) seeks to offer a workable solution to one of metaverse's key issues: using the tactile internet with haptic feedback to allow users in the real world to experience the embodied haptic and kinesthetic feeling of their avatars during interacting with virtual objects or other avatars.

- *Digital twin* allows for real-time replicating of the physical world in various metaverse virtual worlds. Modeling and data fusion are used to achieve this. The metaverse becomes more realistic because of DT, which also makes new service and social interaction facets possible (Bellavista et al., 2022). For instance, Nvidia Omniverse enables BMW to combine its physical auto factories with VR, AI, and robotics to improve its industrial precision and flexibility, which ultimately boosts BMW's planning efficiency by about 30%. For more information, visit https://blogs.nvidia.com/blog/2021/04/13/nvidia-bmw-factory-future/.

- *AI* is a tool that can be used to introduce intelligence into the metaverse for enhanced user experience, such as for effective 3D object rendering, cognitive avatars, and AIGC. For instance, ML is used in EpicGames' MetaHuman project (https://www.unrealengine.com/en-US/digital-humans) to create lifelike digital characters quickly. The produced characters could be used by VSPs to fill the metaverse as talkative virtual assistants provide a thorough analysis of AI training and inference in the metaverse.

- *Economy* sets rules for providing services, trading user-generated content, and providing incentives to support all facets of the metaverse ecosystem. For instance, to synchronize the real and virtual worlds, VSP can pay the price to SSP in exchange for data streams. To accommodate consumers with limited capabilities, metaverse service providers can also buy computational resources from cloud services. Aside from that, the economic system is the motivating force behind the sustainable growth of digital assets and DIDs for the metaverse.

3. **The infrastructure layer** makes it possible to access the metaverse from the edge.
   - *Communication and Networking:* AR/VR and haptic traffic must adhere to strict rate, reliability, and latency standards to avoid breaks in the presence (BIP), or interruptions that make a user aware of the real-world environment. Due to the anticipated enormous expansion in data traffic, ultra-dense networks installed in edge networks may be able to relieve the limited system capacity. Additionally, the B5G/6G communication infrastructure will be crucial in enabling post-Shannon communication to ease bandwidth restrictions and control the skyrocketing development of communication prices (Shen et al., 2021).
   - *Computation:* Modern MMO games can support more than one hundred players concurrently, necessitating high-end GPU needs. The foundational VRMMO games that make up the metaverse system are still hard to come by. The reason is that to render both the immersive virtual worlds and the interactions with hundreds of other players, VRMMO games may require equipment like HMDs to be connected to powerful computers. The cloud edge-end computation paradigm offers a viable approach to enabling universal access to the metaverse. For the least resource-intensive activity, such as the computations needed by the physics engine to determine the movement and positioning of an avatar, local computations can be carried out on edge devices. Edge servers can be used to do expensive foreground rendering, which requires less graphical details but lower latency, to lessen the pressure on the cloud for scalability and further reduce end-to-end latency. The more computationally demanding but less time-critical tasks, such as background rendering, can also be carried out on cloud servers. Additionally, distributed learning and model pruning, and compression AI approaches help to lighten the load on backbone networks.
   - *Blockchain:* The metaverse's economic environment will be established by the DLT that the blockchain provides, which will also be essential for proving ownership of virtual items. Virtual goods today have a hard time finding value outside of the platforms where they are created or traded. To lessen the reliance on such centralization, blockchain technology will be crucial. A non-fungible token (NFT), for instance, authenticates a person's ownership of a virtual asset and serves as a mark of the asset's uniqueness. Peer-to-peer trading is made possible in a decentralized setting by this method, which also safeguards the value of virtual products. The metaverse's virtual worlds are created by many parties, hence the management of user data may also vary. To enable seamless traversal across virtual worlds, multiple parties

will need to access and operate on such user data. Due to value isolation among multiple blockchains, cross-chain is a crucial technology to enable secure data interoperability. In addition, blockchain technology has found recent successes in managing edge resources.

3.6.2.2 Tools, platforms, and frameworks

This section describes several important platforms, frameworks, and tools currently being used in creating the metaverse. AnamXR, Microsoft Mesh, Structure Sensor, Canvas, and Gaimin are a few examples. The characteristics of these development tools that support the metaverse environment are summarized in Table 3.2.

Table 3.2 An overview of the frameworks, platforms, and development tools that support the metaverse ecosystem

Tools	Features
Unity	A tool for producing 2D and 3D content in real-time that incorporates physics, collision detection, and 3D rendering.
Unreal Engine	Photorealistic rendering, dynamic physics and effects, lifelike animation, and reliable data translation are all elements of a real-time 3D production platform.
Roblox	Gives content developers a platform to produce content for the metaverse, much like Unity and Unreal Engine.
Nvidia Omniverse	A 3D simulation and design tool based on RTX technology from Nvidia
Meta Avatars	A platform-agnostic avatar system that can be used to make it easier to create real-world avatars
Meta's Haptics Prototype	A glove that makes it possible for users to sense touches to virtual objects
Hololens2	Mixed reality head-mounted device
Oculus Quest2	VR head-mounted system
Xverse	A blockchain-based virtual environment where users can play, create, share, and work using their digital assets
AnamXR	Unreal engine-based cloud-based virtual e-commerce platform that converts physical goods into pixels
Microsoft Mesh	A system that permits mixed reality applications on any device to provide presence and shared experiences
Structure Sensor	A portable 3D scanner that may be used with mobile devices
Canvas	Software for scanning rooms in 3D and building precise 3D models with Structure Sensor
Gaimin	Permit gaming tokens to be integrated into the metaverse by game and content providers

3.6.3 Communication and ultra-dense networking

Most of the early typical metaverse projects addressed (Diami, 2022; Kevin, 2021; Bernard, 2022) are large-scale endeavours requiring high-end edge devices to function. Therefore, one of the biggest problems with the metaverse is its accessibility. The mobile edge networks are anticipated to offer users high-speed, low-latency, wide-area wireless connectivity for effective communication and networking so they may fully immerse themselves in the metaverse and experience it in real time. To accomplish this, strict communication standards must be met (Table 3.3). The existing 5G network infrastructure is severely taxed by the requirement to transmit 3D virtual objects and scenarios in the metaverse, as opposed to the more conventional transmission of 2D images, to give consumers an immersive experience. To provide immersive content delivery and real-time interaction, enormous networking of metaverse enablers, such as AR/VR, tactile internet, avatars, and DT, requires large bandwidth and URLLC, as shown by the metaverse for mobile edge networks described in (Khan et al., 2022) and ultra-densification of the networking is very much essential. A proposal for a blockchain-based metaverse-native communication system to provide decentralized and anonymous connectivity for all physical and virtual entities concerning encrypted addresses. Real-time execution of production choices and virtual concerts are also required.

Overall, the metaverse's qualities listed below pose serious new obstacles for mobile edge networks while delivering metaverse services:

1. *Immersive 3D streaming:* 3D streaming enables the immersive experience of users in the metaverse to be given to millions of users at mobile edge networks, blurring the line between the real and virtual worlds.
2. *Multi-sensory communication:* The metaverse is a collection of several 3D virtual environments where users can interact as avatars

Table 3.3 Service requirements for communication in the metaverse (Alves et al., 2021)

Applications	Reliability (%)	Latency (ms)	Data rate (Mbit/s)	Connection density (devices/km^2)
The tactile internet	$1-10^{-6}$	1	1	1,000-50,000
A digital twin of smart city	$1-10^{-5}$	5-101	10	100,000
VR entertainment	$1-10^{-5}$	7-15	250	1,000-50,000
Hologram education (point could)	$1-10^{-5}$	20	500-2,000	1,000
AR smart healthcare	$1-10^{-6}$	5	10, 000	50
Hologram education (light field)	$1-10^{-5}$	20	10^5-10^6	1,000

with multi-sensory perception, including vision, hearing, touch, and haptics. Edge users must therefore have a constant connection to the metaverse to use 3D multimedia services like AR/VR and the tactile internet wherever they are.

3. *Real-time interaction:* In the metaverse, user-user and user-avatar real-time interactions are based on massive forms of interactions, such as human-to-human (H2H), human-to-machine (H2M), and machine-to-machine (M2M) communications. Due to the strict requirements for real-time interactions, such as motion-to-photon delay, interaction latency, and haptic perception latency, social multimedia services in the metaverse must be provided.

4. *Seamless physical-virtual synchronization:* To offer services for the synchronization of the physical and virtual worlds, physical and virtual entities must interact and exchange real-time status updates. The age of information (AoI) or value of information (VoI) may have an impact on the value and importance of synchronizing data.

5. *Multi-dimensional collaboration:* To maintain the metaverse, VSPs, and PSPs in both the real and virtual worlds must work together multi-dimensionally (Pan et al., 2021). For instance, as was covered in Section 3.6.2.1, the physical entities in edge networks, such as sensor networks run by PSPs, must regularly gather data from the outside world to maintain the currency of that data for P2V synchronization. Decisions made in the virtual world can then be translated into actions in the real world. This loop necessitates the cooperation of numerous real-world and virtual entities across several dimensions, such as time and space.

A description of the networking and communication possibilities available to edge-enabled devices at the cutting-edge metaverse is thought to help with these issues. Users must first become fully immersed in the metaverse through the seamless delivery of 3D multimedia services in Section 3.6.3.1. This includes assistance with VR streaming and AR adaptation, enabling users and real-world objects to synchronize with virtual worlds without any difficulty. In addition to conventional content delivery networks, communication, and networking that underpins the metaverse must give user-centric concerns precedence in the content delivery process, as indicated in Section 3.6.3.2. The metaverse is chock full of contextual and tailored content-based services, like augmented reality and virtual reality (AR/VR) and tactile internet. Additionally, the rapid expansion of data and constrained bandwidth need a paradigm shift away from the traditional concentration of classical. In response, evaluating semantic/goal-oriented communication strategies can assist future multimedia services in making better use of the scarce spectrum. Real-time bidirectional physical-virtual synchronization, or DT, is essential for the creation of the metaverse, which is addressed in Section 3.6.3.3. UAVs and reconfigurable intelligent surfaces

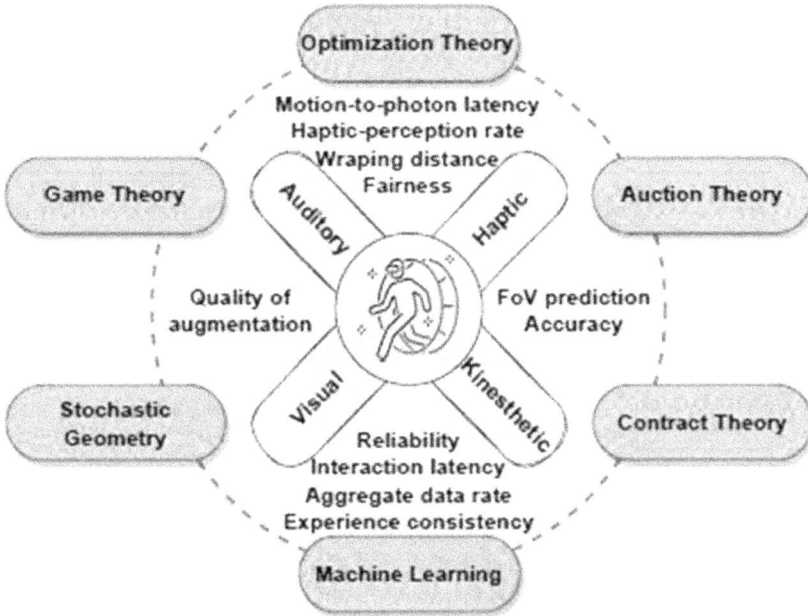

Figure 3.6 Multisensory multimedia networks designed using mathematical methods and metrics that focus on people.

(RIS) will be used to simplify the sensing and actuation interaction between the physical and digital worlds. Table V of Xu et al. (2023) gives a summary of the reviewed works in terms of situations, issues, performance measures, and mathematical tools. Figure 3.6 also shows the analytical mathematical tools and measurements that focus on people.

3.6.3.1 Rate-reliability-latency 3D multimedia networks

The seamless and immersive experience of embodied telepresence via AR/VR brought by the metaverse places high demands on the communication and network infrastructure of mobile edge networks in terms of transmission rate, reliability, and latency. In particular, the high volume of data exchanged to support AR/VR services that traverse between the virtual worlds and the physical world requires the holistic performance of the edge networking and communication infrastructure that optimizes the trade-off among rate, reliability, and latency. The first dimension of consideration is the data rate that supports round-trip interactions between the metaverse and the physical world (e.g., for users and sensor networks aiding in the physical-virtual world synchronization) (Hieu, 2022). Second, interaction latency is another critical challenge for users to experience realizm in AR/VR services. Mobile edge networks, for example, allow players to ubiquitously participate in massively multiplayer online games that require ultra-low latency for smooth interaction. The reason is that latency determines how quickly players receive information about their situation in the virtual worlds and how quickly their responses are transmitted to other players. Finally, the third dimension is the reliability of physical network services, which refers to the frequency of BIP for users connecting to the metaverse. Moreover, the reliability requirements of AR/VR applications and users may vary dynamically over time, thereby complicating the edge resource allocation for AR/VR services provision of VSPs and PSPs.

3.6.3.2 Human-in-the-loop communication

A 3D virtual environment focused on humans and supported by extensive machine-type communication is called the metaverse. These human-centered applications, such as holographic augmented reality social media, virtual reality games, and tactile internet-enabled healthcare, will be the driving forces behind wireless networks in the future. It is necessary to integrate not only engineering for service supply (communication, processing, storage), but also interdisciplinary concerns of human body perception, cognition, and physiology for haptic communication, to create highly realiztic XR systems. The term "haptics" refers to both kinesthetic perception (the knowledge of forces, torques, position, and velocity experienced by muscles, joints, and tendons of the human body) and tactile perception (the perception of surface roughness and friction sensed by many types of mechanoreceptors). Some additional measures, such as quality of physical experience (QoPE) are proposed to assess the calibre of human-in-the-loop communication to further assess the effectiveness of the service providers in the metaverse. These measures tend to combine the traditional inputs of QoS (latency, rate, and dependability) and QoE with physical aspects of human users, such as their physiological and psychological perception (average opinion score). Different human-related activities, such as cognition, physiology, and gestures, have an impact on the parameters impacting QoPE.

3.6.3.3 Real-time physical-virtual synchronization

The line separating the real and virtual worlds is blurred by the metaverse, a novel medium (Duan et al., 2021). The metaverse, for instance, can enhance the performance of mobile edge networks through offline simulation and online decision-making by preserving a digital representation of the actual wireless environment in the virtual worlds (Pan et al., 2021). To enable working from home in the metaverse, for instance, staff members can use DTs to create digital replicas of their actual workplaces. For sensing and actuation, widely dispersed edge devices and edge servers are necessary to maintain bidirectional real-time synchronization between the metaverse

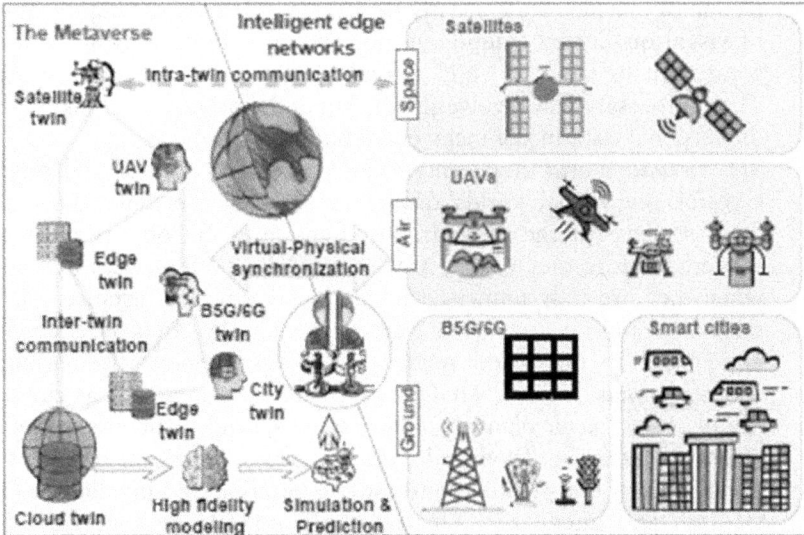

Figure 3.7 Metaverse and intelligent edge networks synchronizing in real time between the physical and virtual worlds.

and the actual world. The DT which refers to digital replicas of actual physical entities in the metaverse, such as twin cities and digital versions of smart factories, is one of the potential solutions for such virtual services, as depicted in Figure 3.7. The DT can offer high-quality modelling, simulation, and prediction for physical entities based on historical data from digital replications. The metaverse can be utilized to increase the effectiveness of the edge networks through intra-twin and inter-twin communication. Users can rapidly control and calibrate their physical entities in the real world through the metaverse by specifically monitoring the DTs of edge devices and edge infrastructure like RIS, UAVs, and space-air-ground integrated network SAGIN.

3.6.4 Computation

The resource-intensive services offered by the metaverse will want expensive processing resources in addition to dependable and low-latency communication networks (Duan et al., 2021). The following calculations must be carried out to use the metaverse:

- *High-dimensional Data Processing:* To enable users to experience realizm in the metaverse, the physical and virtual worlds will produce enormous amounts of high-dimensional data, such as spatiotemporal data. For instance, for realizm, a virtual object that has been dropped by a user should behave by physical rules. Real-time ray tracing technology, which was recently developed to deliver the highest level of visual quality by computing more bounces in each light ray in a scene, can be used to build virtual worlds with physical features. These processes will involve processing and storing the created high-dimensional data in the metaverse's databases.

- *3D Virtual World Rendering:* The process of rendering involves transforming virtual worlds' raw data into 3D objects that can be seen on a screen. All the photographs, films, and 3D objects must be rendered before they can be viewed on AR or VR devices by users when they are fully immersed in the metaverse. For instance, Meta runs museums to convert 2D artwork into 3D. Metaverse is also anticipated to build and render 3D things intelligently. Through acoustic inputs, AI (i.e., AIGC) may create 3D things such as clouds, islands, trees, picnic blankets, tables, stereos, drinks, and even sounds.
- *Avatar Computing:* In virtual worlds, avatars are digital representations of users that require constant computation and intelligence for creation and interaction. On the one hand, ML technologies like computer vision (CV) and natural language processing are the foundation for avatar production (NLP). On the other hand, analyzing and forecasting the outcomes of the real-time interaction between avatars and users requires a lot of processing power. AI and AR applications, for instance, can work together to offer avatar services like obtaining text information from the users' field of view, word amplification on paper text, and copying and pasting files between computers with a hand gesture.

Users at mobile edge networks should be able to access the metaverse anywhere. As a result, processing resources are needed by both users' edge devices to access the metaverse at mobile edge networks as well as by metaverse service providers, such as for user analytics. For instance, an AR/VR recommender system can be taught to suggest immersive material that pleases viewers using data from mobile devices (Lam, 2021). Second, AR/VR rendering demanded an immersive experience from consumers. The main computation considerations are listed below:

- *Ubiquitous computing and intelligence:* Services are needed to surround and support users of diverse resource-constrained devices who can access the metaverse at any time and from any location (Huynh-The et al., 2023). Users can, on the one hand, delegate computationally demanding jobs to edge computing services to reduce straggler effects and boost performance. On the other side, stochastic demand and user mobility need the provision of continuous and adaptive computing, as well as intelligence services, at mobile edge networks.
- *Embodied telepresence:* Without being constrained by physical distance, users from many regions can be telepresent at the same

time in the metaverse's virtual worlds (Lee et al., 2021). The metaverse is anticipated to display 3D objects in AR/VR effectively using ubiquitous computation resources at mobile edge networks to provide users with an embodied telepresence.

- *Personalized AR/VR recommendation:* Users in the metaverse have personal preferences for various types of augmented reality and virtual reality material (Duan et al., 2021). Therefore, depending on consumers' history and contextual data, content providers must generate individualized suggestions for AR/VR content. The QoE of users is enhanced in this way.

- *Cognitive Avatars:* Users navigate the metaverse by controlling their cognitively capable avatars. The AI-powered cognitive avatars' (Kocur et al., 2020) ability to mimic their owners' behaviour and behaviours allows them to move and behave naturally in the metaverse. Users can manage several avatars at once in various virtual worlds in this way. Additionally, users can avoid some light but laborious activities that cognitive avatars can handle.

- *Privacy and security:* For a variety of computation tasks, such as rendering AR/VR and creating avatars, the metaverse will be supported by ubiquitous computing servers and devices (Pietro & Cresci, 2021). However, the reliance on the distributed computing paradigm, such as mobile cloud computing, suggests that users would need to share their data with outside parties. The metaverse will gather copious amounts of data about users' physiological reactions and bodily motions, including sensitive and private information like habits and physiological traits. To prevent data from being leaked to outside parties when a user is immersed in the metaverse, privacy, and security in the computation are essential factors to take into account.

The cloud-edge-end computation model is briefly described in Section 3.6.4.1. The computing techniques for resource-efficient AR/VR cloud edge rendering are then covered in Section 3.6.4.2.

3.6.4.1 The paradigm of cloud-edge end computing

This section describes how the mobile cloud-edge end collaborative computing paradigm enables ubiquitous computing and intelligence for users and service providers in the metaverse, as shown in Figure 3.8.

- *Mobile cloud computing:* Numerous tech behemoths have set the full dive metaverse as their most recent macro-goal. Billions of dollars will be invested in cloud gaming over the next ten years with the idea that these technologies will support our online-offline virtual future. High-quality video games can be played via cloud gaming, which integrates cloud computing. Mobile devices or mobile embedded systems can use

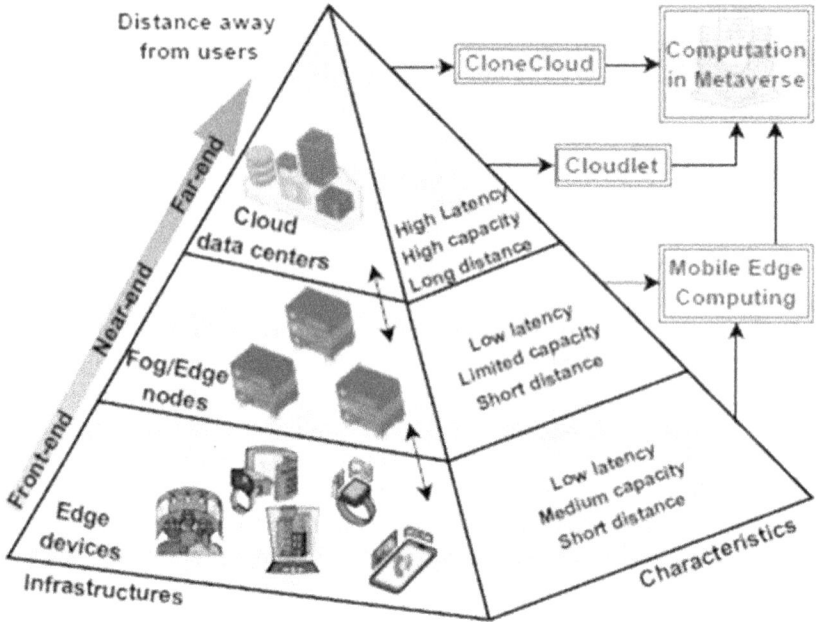

Figure 3.8 The properties of several forms of computer infrastructure to facilitate computation in the metaverse.

mobile cloud computing to access computing services. Virtual machines are used by the cloud provider to deliver services to customers while installing compute servers in their infrastructure. The most economical strategy for a business, though, is to integrate internal and cloud resources rather than favouring one over the other. Fortunately, a few system architectures have been developed to manage mobile cloud computing and offer support for devices with weak computation so that they can access the metaverse, such as Cloudlet and CloneCloud.

- *Edge computing:* The metaverse is a computational resource "black hole." Delivering real-time, sophisticated, computationally intensive features like speech recognition and AR/VR will always demand the highest feasible compute capacity, and mobile users accessing the metaverse will need the latency to be as low as possible. To maintain the metaverse, edge computing offers minimal latency, great efficiency, and security. Fog computing and edge computing share a similar idea. Edge and fog computing both keep track of the local network's capacity for computation and compute tasks that would have been better off being handled in the cloud, decreasing latency and the strain on backbone networks. Edge computing, on the other hand, makes use of the extra computational power and storage offered by edge devices. When placed between the edge and the nodes that produce and analyse IoT data, the fog layer can be considered an extension of the mobile cloud.

- *Fog Computing:* Billions of users and IoT devices will have constant communication thanks to the metaverse. A never-before-seen creation of massive volumes and varied modalities of data will result from this. Due to the inescapable transmission latency to far-off cloud servers, mobile cloud computing can no longer satisfy the high expectations for real-time interaction of immersive and social apps. To bring cloud services closer to customers and decrease latency when computation duties are offloaded, fog computing was first proposed. The authors (Sun et al., 2022) provide theoretical evidence demonstrating that fog computing has lower latency than conventional cloud computing. Fog computing, in contrast to mobile cloud computing, deploys several fog nodes between the cloud and edge devices (Tan et al., 2020). For instance, because they are unable to establish a reliable connection with the cloud, mobile users who commute by train are unable to access the metaverse. Therefore, the locomotive can be equipped with fog computing nodes to enhance communication. Industrial gateways, routers, and other equipment with the required computing power, storage capacity, and network connection could be fog nodes. Fog nodes can connect edge devices and users via wireless connection methods like B5G and 6G while also offering computing and storage services to reduce latency.

3.6.4.2 Effective cloud-edge end rendering for AR and VR

Various mobile and wired devices, such as HMD and AR goggles, can access the edge-enabled metaverse. VR enables users to experience a simulated virtual world while in the real world. The virtual worlds are created by VR equipment using sensory images (Duan et al., 2021). In the metaverse, real-time rendering is necessary so that the sensory images may be created swiftly enough to constitute a continuous flow as opposed to discrete events. For instance, educational systems can use VR to enhance how lessons are delivered in virtual environments. In contrast to VR, AR enables interaction with the real environment that has been digitally enhanced or modified in some way. For situations when users are present in the real environment, such as on-the-job training and computer-assisted work, AR apps are ideally suited. However, the computing power, memory, and battery of consumers' devices are constrained. Such devices cannot support intensive AR/VR apps needed for the immersive metaverse. To solve these difficulties, users' devices can take advantage of ubiquitous computing resources and intelligence for remote rendering and task offloading based on the cloud-edge end collaborative computing model.

The cloud-edge end computing architecture for mobile edge networks enables the execution of computations such as high-dimensional data processing, 3D virtual environment rendering, and avatar computing at the source of the data. For instance, the car might deliver the data to neighbouring vehicles or roadside units to execute the computing instead of offloading it to the cloud, drastically reducing end-to-end latency when the user interacts with avatars in the metaverse from the vehicle. By utilizing the computing resources at the mobile edge networks, the suggested cloud edge-end collaborative computing paradigm is a viable way to reduce latency for mobile users. This is crucial because mobile users should have access to metaverse whenever and wherever they are. Additionally, by moving computation from the internet to the edge, mobile edge computing reduces network traffic. Edge servers at mobile edge networks are outfitted with edge servers that can give mobile users access to computing resources. When deciding whether mobile users should execute binary offloading to edge servers or minimal offloading, for instance, divide a task into smaller sub-tasks. Thus, to enhance user QoS, VR content can be sent to neighboring edge servers for real-time rendering. The most common benefit of the systems evaluated, as can be seen, is their ability to adapt to diverse settings and requirements of AR/VR rendering. However, they have a few drawbacks, such as the possibility of being partial, unjust, impractical, or erroneous solutions.

3.6.5 Directions for future research in the metaverse

When different technologies are adopted in the metaverse, this section covers the major future paths.

- *Immersive Streaming with Advanced Multiple Access:* The provisioning of unprecedented ubiquitous accessibility, extremely high data rates, and URLLCs of the mobile edge networks is required by the heterogeneous services and applications in the metaverse, such as AR/VR, the tactile internet, and hologram streaming. To share the virtual worlds most efficiently, the shared metaverse specifically requires complex multiple-access techniques to support numerous users in the allowed resource blocks, such as time, frequency, codes, and power.
- *Networks for Multisensory Multimedia:* The metaverse offers customers multi-sensory multimedia services, such as AR/VR, the tactile internet, and holographic streaming, in contrast to the conventional 2D internet. Mobile edge networks must be able to simultaneously offer holographic services (such as AR/VR and the tactile internet) to support these multi-sensory multimedia services. However, because multiple types of network resources are needed at once, it is impossible to build resource allocation algorithms for such intricate multi-sensory multimedia services. For instance, URLLC services are needed for the tactile internet whereas eMMB services are needed for AR/VR. To be more specific, while VR services frequently use the network's downlink

transmission resources and caching capabilities, augmented reality services frequently use its uplink transmission resources and computation resources. Therefore, effective and suitable resource allocation systems should be proposed in these multi-sensory multimedia networks to enable the user's immersive experience.

- *Semantic/Goal-Aware Multimodal Communication:* Semantic communication transforms mobile edge networks from data-oriented to semantic-oriented, supporting a wide range of context-aware and goal-driven services in the metaverse. This is accomplished by automatically understanding AI models. Existing semantic communication models, on the other hand, frequently concentrate on semantic extraction, encoding, and decoding for a single activity, such as speech or images. Service models in the metaverse typically incorporate a variety of instantaneous interactions, including voice and video services. This brings up new problems for semantic communication, such as (i) how to create a multimodal semantic communication model to deliver multi-sensory multimedia services in the metaverse; (ii) how to effectively extract semantics from the data transmitted by users; and (iii) how to allocate resources in the edge network to support the training and use of semantic communication models.

- *Communication and Sensing Integrated:* Beginning with the creation of the metaverse (through physical-virtual synchronization) and continuing through user access, communication, and sensing are pervasive in the edge network. Future mobile edge networks, however, will also utilize the spectrum often reserved for sensing (such as mm wave, THz, and visible light). To assure the creation and maintenance of the metaverse, it is imperative that communication and sensing are integrated. For real-world physical entities that contain a lot of static and dynamic information, a DT, for instance, constructed the metaverse. Additionally, the metaverse must allow for seamless communication and interaction between users and entities, as well as instantaneous distribution of such information to targeted users. In general, better sensor and communication integration enables more complete metaverse services and real-time digital replication of the physical environment.

- *Digital Twin Edge Networks:* The metaverse's most efficient engine for synchronizing the real-time states of the physical and virtual worlds is called DT. With the aid of a huge number of edge nodes and IoE devices, the DT may digitally replicate, monitor, and manage real-world entities at mobile edge networks. Furthermore, by connecting digital entities in the metaverse, entities in edge networks can function in the real world more effectively. Although DT can improve mobile edge network operation and maintenance, they also demand substantial communication, computing, and storage resources, making them challenging for edge networks with limited resources to handle. Therefore, a focused study area will be on more effective DT solutions and better DT-enabled mobile edge network operations and maintenance.

- *Intelligent Edges and Edge Intelligence:* Future research should concentrate on the edge intelligence-driven infrastructure layer, which is an essential component of the future wireless networks, to actualize the metaverse despite its particular challenges. Edge computing and AI are essentially merging to become edge intelligence. The two key aspects of edge intelligence, which are edge for AI and AI for edge, should be implemented. Edge for AI refers to an end-to-end framework that brings sensing, communication, training of AI models, and inference closer to the location where data is produced. AI for edge is the application of AI algorithms to enhance the orchestration of the aforementioned framework.

- *Allocation of Resources Sustainably:* The metaverse will always have a resource problem. For mobile devices or portable technology with limited storage and processing power to offer good and immersive QoE to mobile consumers, cloud, fog, and edge computing services are always required. As a result, the metaverse will require a growing amount of energy to support communication and processing, which will increase energy consumption and raise environmental issues like greenhouse gas emissions. Therefore, to accomplish sustainability, green cloud, fog, and edge networking and computing are required. A few potential solutions include developing new architectures with sustainability in mind to support green cloud/fog/edge networking and computing in the metaverse; ii) allocating resources in an energy-efficient way; and iii) fusing green cloud/fog/edge networking and computing with other emerging green technologies.

- *Digital Human Avatars:* Users of the metaverse are telepresent and fully submerged in the virtual worlds as avatars or digital beings. An essential point to ask is how to set up dynamic avatar services at mobile edge networks to improve the QoE of avatar services, as human-like avatars may need a lot of resources to build and maintain. Additionally, each avatar must gather and retain a sizable amount of personal biological information from the user or other users who have mapping relationships with the avatar in the virtual worlds or associated physical entities in the real world to provide avatar services. Therefore, it is important to maintain privacy and secure data during the creation and use of avatars.

- *Blockchain with Intelligence:* Intelligent blockchain uses AI algorithms to make traditional blockchain operations like consensus and block propagation more adaptable to the current network environment. The convergence of AI and blockchain, which together make up the metaverse's fundamental engine and infrastructure, enables the metaverse to effectively safeguard users' security and privacy. There have been some preliminary investigations into the intelligent blockchain, including work on dynamic routing algorithms for payment channel networks and adaptive block propagation models. The current study on the intelligent blockchain, however, solely aims to improve the blockchain's performance in virtual settings. The operation of physical entities (such as wireless BSs, automobiles, and UAVs) in the physical world will change as intelligent blockchain is connected to the physical world in the metaverse.

- *Quality of Experience:* While immersed in the virtual worlds and engaging with other users, individuals, and their avatars should have a satisfying human-felt QoE. To rate and manage the supplied services in the metaverse, both subjective and objective QoE measurements based on physiological and psychological studies can be used.
- *Design of the Metaverse Services Market and Mechanisms:* Innovative market and mechanism design is essential for interactive and resource-intensive services in the metaverse to make it easier for service providers and customers to allocate resources and set prices. The market and mechanism design for metaverse services should take local states of the physical and virtual submarkets while taking into account their interplay effects of them because the metaverse has the potential to blur the line between the physical and virtual worlds.
- *The Industrial/Automotive Metaverse:* For next-generation intelligent production, the burgeoning industrial metaverse will connect real factories with virtual worlds. The Industrial Internet of Things (IIoT) allows the industrial metaverse to collect data from various production and operation lines, which allows it to conduct efficient data analysis and decision-making, improving physical space production efficiency while lowering operating costs and increasing commercial value. The vehicular metaverse, however, incorporates immersive streaming and real-time synchronization, which is anticipated to improve driving efficiency, safety, and the immersive experience for drivers and passengers.

Three factors, embodied user experience, harmonious and sustainable edge networks, and comprehensive edge intelligence, will guide future research on the metaverse at mobile edge networks. First, the delivery of services through a 3D embodied internet in the metaverse that is immersive and human-aware will encourage the development of multi-sensory multimedia networks and human-in-the-loop communication. The advent of the metaverse, on the other hand, motivates mobile edge networks to offer resilient computing and communication services via real-time physical-virtual synchronization and reciprocal optimization in digital edge twin networks. Third, environments in physical and virtual worlds can interact in the metaverse without posing a security or privacy risk because of extensive edge intelligence and blockchain services.

3.7 CONCLUSIONS

3.7.1 Ultra-densification of networks

The pros and cons of network densification are covered in this chapter. The BS and ED components of modern network densification are outlined. The main difficulties in BS densification, such as interference control, CoMP design, and BS activation control, are then discussed. There is also a list of the main difficulties the ED densification has encountered, such as problems with pilot contamination, user activity detection, and BS association. To overcome these difficulties, the main advantages of network densification in the 6G wireless communication network, including functionalized BSs, spatial correlation exploitation, and intellectualized buffer strategy, are also offered. By considering ED and BS densification, this chapter offers major design insights for the upcoming multi-tier dense networks.

3.7.2 Metaverse

This study starts by giving a general overview of the metaverse's architecture, as well as an analysis of its present state of development and related technologies, where communication networking densification is one of the major enablers. Next, with a focus on the edge-enabled metaverse, it is discussed how crucial it is to resolve significant networking, processing, and communication problems. The ultimate goal of the metaverse is described, which sheds light on potential routes for further study in this field. Before a thorough analysis of the metaverse at edge networks, the survey is the first stage. It provides knowledge and assistance to academics and practitioners for further work on the edge-enabled metaverse.

REFERENCES

Alves, H., Jo, G. D., Shin, J., Yeh, C., Mahmood, N. H., Lima, C., ... Latva-aho, M. (2021). Beyond 5G URLLC evolution: New service modes and practical considerations. Retrieved from http://arxiv.org/abs/2106.11825

Alwis, C. D., Kalla, A., Pham, Q. V., Kumar, P., Dev, K., Hwang, W. J., & Liyanage, M. (2021). Survey on 6G frontiers: Trends, applications, requirements, technologies and future research. *IEEE Open Journal of the Communications Society*, 2, 836–886. 10.1109/ojcoms.2021.3071496

Bates, A. E., Primack, R. B., Biggar, B. S., Bird, T. J., Clinton, M. E., Command, R. J., & Duarte, C. M. (2021). Global COVID-19 lockdown highlights humans as both threats and custodians of the environment. *Biological Conservation*, 263(109175), 109175. doi:10.1016/j.biocon.2021.109175

Bellavista, P., Giannelli, C., Mamei, M., Mendula, M., & Picone, M. (2022, September). Digital twin oriented architecture for secure and QoS aware intelligent communications in industrial environments. *Pervasive and Mobile Computing*, 85, 101646. 10.1016/j.pmcj.2022.101646.

Bernard, M. (2022). The amazing possibilities of healthcare in the metaverse. Accessed Feb. 16 2022, https://www.forbes.com/sites/bernardmarr/2022/02/23/the-amazing-possibilities-of-healthcare-in-the-metaverse/

Chang, L., Zhang, Z., Li, P., Xi, S., Guo, W., Shen, Y., Xiong, Z., Kang, J., Niyato, D., Qiao, X., & Wu, Y. (2022, June). 6G-enabled edge AI for metaverse: Challenges, methods, and future research directions. *Journal of Communications and Information Networks*, 7(2), 107–121. 10.23919/jcin.2022.9815195

Diami, V. (2022). What comparisons between second life and the metaverse miss. Accessed Feb. 16, 2022, https://slate.com/technology/2022/02/second-life-metaverse-facebook-comparisons.html.

Duan, H., Li, J., Fan, S., Lin, Z., Wu, X., & Cai, W. (2021). Metaverse for social good: A university campus prototype. Retrieved from http://arxiv.org/abs/2108.08985.

Gupta, A. K., Sabu, N. V., & Dhillon, H. S. (2021). Fundamentals of network densification. In: Lin, X. & Lee, N. (eds), *5G and Beyond*. Springer, Cham. 10.1007/978-3-030-58197-8_5

Hieu, N. Q., Nguyen, D. N., Hoang, D. T., & Dutkiewicz, E. (2022). When virtual reality meets rate splitting multiple access: A joint communication and computation approach. Retrieved from http://arxiv.org/abs/2207.12114

Huynh-The, T., Gadekallu, T. R., Wang, W., Yenduri, G., Ranaweera, P., Pham, Q. V., da Costa, D. B., & Liyanage, M. (2023, June). Blockchain for the metaverse: A review. *Future Generation Computer Systems*, 143, 401–419. 10.1016/j.future.2023.02.008.

Jeong, H., Yi, Y., & Kim, D. (2022). An innovative e-commerce platform incorporating metaverse to live commerce. *International Journal of Innovative Computing, Information and Control*, 18(1), 221–229. 10.24507/ijicic.18.01.221

Jiang, M., Cezanne, J., Sampath, A., Shental, O., Wu, Q., Koymen, O., Bedewy, A., & Li, J. (2022, April). Wireless fronthaul for 5G and future radio access networks: Challenges and enabling technologies. *IEEE Wireless Communications*, 29(2), 108–114. 10.1109/mwc.003.2100482.

Kevin, S. (2021). Nvidia CEO says the metaverse could save companies billions of dollars in the real world. Accessed Nov. 25, 2021, https://www.cnbc.com/2021/11/19/nvidia-ceo-says-the-metaverse-could-save-companies-billions.html

Khan, L. U., Han, Z., Niyato, D., Hossain, E., & Hong, C. S. (2022). Metaverse for wireless systems: Vision, enablers, architecture, and future directions. Retrieved from http://arxiv.org/abs/2207.00413.

Kocur, M., Schauhuber, P., Schwind, V., Wolff, C., & Henze, N. (2020). The effects of self-and external perception of avatars on cognitive task performance in virtual reality. In Proceedings of the 26th ACM Symposium on Virtual Reality Software and Technology, Virtual Event (pp. 1–11).

Lam, K. Y., Lee, L. H., & Hui, P. (2021, October 17). A2W: Context-aware recommendation system for mobile augmented reality web browser. In Proceedings of the 29th ACM International Conference on Multimedia. Presented at the MM '21: ACM Multimedia Conference, Virtual Event China. 10.1145/3474085.3475413.

Lee, Lik-Hang, Braud, Tristan, Zhou, Pengyuan, Wang, Lin, Xu, Dianlei, Lin, Zijun, Kumar, Abhishek, Bermejo, Carlos, & Hui, Pan. (2021). All one needs to know about metaverse: A complete survey on technological singularity, virtual ecosystem, and research agenda. 10.13140/RG.2.2.11200.05124/8.

Lim, W. Y. B., Xiong, Z., Niyato, D., Cao, X., Miao, C., Sun, S., & Yang, Q. (2022). Realizing the metaverse with edge intelligence: A match made in heaven. *IEEE Wireless Communications*, 30(4), 64–71, 10.1109/MWC.018.2100716.

Mahmood, Nurul, Alves, Hirley, Alcaraz López, Onel, Shehab, Mohammad, Moya Osorio, Diana Pamela, & Latva-aho, Matti. (2020). Six Key features of machine type communication in 6G. 1–5. 10.1109/6GSUMMIT49458.2020.9083794.

Mukherjee, S., Kim, D., & Lee, J. (2021, November). Base station coordination scheme for multi-tier ultra-dense networks. *IEEE Transactions on Wireless Communications*, 20(11), 7317–7332. 10.1109/twc.2021.3082625.

Pan, J., Cai, L., Yan, S., & Shen, X. S. (2021, November). Network for AI and AI for network: Challenges and opportunities for learning-oriented networks. *IEEE Network*, 35(6), 270–277. 10.1109/mnet.101.2100118.

Park, C., Mukherjee, S., & Lee, J. (2021). Base station activation in coordinated multi-point joint transmission-based ultra dense networks. In 2021 IEEE Globecom Workshops, GC Wkshps 2021-Proceedings. Institute of Electrical and Electronics Engineers Inc.

Pietro, D., & Cresci, S. (2021). Metaverse: Security and privacy issues. In 2021 Third IEEE International Conference on Trust, Privacy and Security in Intelligent Systems and Applications (TPS-ISA) (pp. 281–288). Atlanta, GA.

Polese, M., Giordani, M., Zugno, T., Roy, A., Goyal, S., Castor, D., & Zorzi, M. (2020, March). Integrated access and backhaul in 5G mmWave networks: Potential and challenges. *IEEE Communications Magazine*, 58(3), 62–68. 10.1109/mcom.001.1900346.

Ray, P. P., Kumar, N., & Guizani, M. (2021, August). A vision on 6G-enabled NIB: Requirements, technologies, deployments, and prospects. *IEEE Wireless Communications*, 28(4), 120–127. 10.1109/mwc.001.2000384

Shen, X., Gao, J., Wu, W., Li, M., Zhou, C., & Zhuang, W. (2021). Holistic network virtualization and pervasive network intelligence for 6G. *IEEE Communications Surveys & Tutorials*, 24(1), 1–30.

Sun, Yaohua, Chen, Jianmin, Wang, Zeyu, Peng, Mugen, & Mao, Shiwen. (2022). Enabling mobile virtual reality with open 5G, fog computing and reinforcement learning. *IEEE Network*, 36(6), 142–149. 10.1109/MNET.010.2100481.

Tan, Z., Qu, H., Zhao, J., Zhou, S., & Wang, W. (2020). UAV-aided edge/fog computing in smart IoT community for social augmented reality. *IEEE Internet of Things Journal*, 7(6), 4872–4884. 10.1109/jiot.2020.2971325

Tang, F., Chen, X., Zhao, M., & Kato, N. (2022). The roadmap of communication and networking in 6G for the metaverse. *IEEE Wireless Communications*, 1, 1–15. 10.1109/MWC.019.2100721.

Wu, Y., Zhang, K., & Zhang, Y. (2021). Digital twin networks: A survey. *IEEE Internet of Things Journal*, 8(18), 13789–13804. doi:10.1109/jiot.2021. 3079510

Xu, M., Ng, W. C., Lim, W. Y. B., Kang, J., Xiong, Z., Niyato, D., Yang, Q., Shen, X., & Miao, C. (2023). A full dive into realizing the edge-enabled metaverse: Visions, enabling technologies, and challenges. *IEEE Communications Surveys & Tutorials*, 25(1), 656–700. 10.1109/comst.2022.3221119

Yaacoub, E., & Alouini, M. S. (2020, April). A key 6G challenge and opportunity-connecting the base of the pyramid: A survey on rural connectivity. *Proceedings of the IEEE*, 108(4), 533–582. 10.1109/jproc.2020.2976703.

Yuan, Y., Mann, S., Furness, T., Rosedale, P., Trevett, N., Lebaredian, R., Kalinowski, C., Lange, D., Miralles, E., Inbar, O., & Ball, M. (2022). Metaverse landscape & outlook: Metaverse decoded by top experts. *IEEE Metaverse Congress*, https://lnkd.in/eWC36DuV.

Zhan, J., & Dong, X. (2021, March). Interference cancellation aided hybrid beamforming for mmWave multi-user massive MIMO systems. *IEEE Transactions on Vehicular Technology*, 70(3), 2322–2336. 10.1109/tvt.2 021.3057547.

Zhou, Y., Xiao, X., Chen, G., Zhao, X., & Chen, J. (2022). Self-powered sensing technologies for human metaverse interfacing. *Joule*, 6, 1368–1389. 10.1016/ j.joule.2022.06.011.06.011.

Chapter 4

Cognitive radios

N. Chitra Kiran

Department of ECE, Alliance University, Bangalore, India

4.1 COGNITIVE RADIOS

A transceiver in a cognitive radio (CR) system may intelligently distinguish between channels that are actively being used and those that are not. The transceiver immediately switches to unoccupied channels and avoids busy ones. These features allow for more efficient utilization of the radio spectrum. As a bonus, it reduces the likelihood of causing problems for other users. Furthermore, it enhances spectrum efficiency and user satisfaction by avoiding congested channels. Because of its scarcity, wireless RF spectrum is typically assigned through a licensing system. The Federal Communications Commission (FCC) and the National Telecommunications and Information Administration (NTIA) are responsible for this in the United States. In the United States, the FCC is responsible for managing the airwaves for commercial and industrial usage, while the NTIA handles government use. Spectrum allocations are not always used efficiently. As a result, some bands are crammed while others go underutilized. As a result, the amount of data that can be sent to customers and the quality of the service provided are both diminished by this inefficient use of the spectrum. CR is comprised of two primary systems: a decision-making cognitive unit and a software-defined radio (SDR) unit that can function in several different modes. When designing a CR, it is common practice to also have a dedicated spectrum sensing (SS) subsystem for gathering environmental signal data and identifying co-existing services and users. It's worth noting that these subsystems don't always correspond to a single piece of hardware but rather may include parts that are dispersed across a larger network. Therefore, it is common to speak to a CR system or a cognitive network when discussing CR. Figure 4.1 depicts spectral inefficiency, with certain spectrum bands congested while others are underutilized.

This restricted resource is getting increasingly scarce as the number of connected devices in use increases. To make the most of and divide up this scarce resource, CR can be put to good use. As wireless technology has improved, so too has the prevalence of wireless gadgets in modern society.

DOI: 10.1201/9781003369028-4

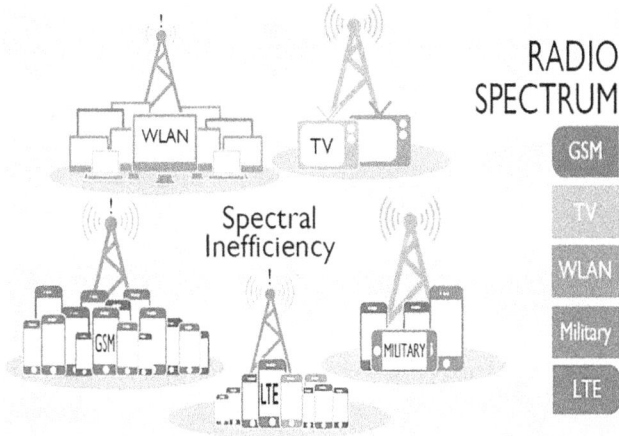

Figure 4.1 Spectral inefficiency illustration.

As IoT becomes more widely adopted, we may anticipate a dramatic increase in the number of linked devices. An enormous amount of radio spectrum is needed to accommodate the proliferation of wireless gadgets. However, usable spectrum is a limited resource. Looking at the present spectrum allocation chart, it becomes clear that there will not be enough available spectrum to accommodate the expected growth in wireless devices and mobile data traffic shortly. As a solution to the impending spectrum crunch problem, the idea of "CR" has been suggested. Unlicensed users of CR can dynamically locate and use empty licensed spectrum without interfering with legitimate users. Technologies including SS, a spectrum database, and a pilot channel are already in use for CR. These methods are either overly complicated to implement and so demand a lot of computing effort to discover unused spectrum or they do not make use of newly created spectrum space in real-time. To accommodate the astronomical growth in mobile data usage in forthcoming wireless networks, two promising new technologies are emerging: sixth-generation (6G) wireless networking and CR. It is expected that the 6G cognitive radio networks (CRNs), with their high-performance matrices, will outperform their 5G. While 5G's peak data rate is 20 Gbps, 6G's with the help of the optical frequency band and THz is predicted to reach 1–10 Tbps. With such maximum frequency bands, the data rate used by users can increase to a Gbps level. Zone-specific traffic capacities could exceed 1 Gbps/m^2. Spectrum efficiency has the potential to improve by a factor of three to five, and energy efficiency should improve by a factor of many orders of magnitude (about 100) compared to 5G. Applying AI to network automation and management brings us one step closer to this goal. Due to massively heterogeneous networks, wide bandwidth, enormous antennas, and various communication conditions, the connection density will rise by a factor of around a thousand. Security capability, cost-effectiveness, intelligence, and capacity are also important matrices. A human-centric CRN communication protocol in 2030 is primarily supported by implantable sensors, wearable device communication, smart industrial 4.0, intelligent vehicles, and robotics.

CR was first proposed in 1998 by Joseph Mitola of Stockholm's KTH Royal Institute of Technology. Software-defined radio combined with spread spectrum transmission forms a hybrid technology. The primary network and the secondary network are the two halves of a CRN. Those who utilize the principal radio base station and are part of the primary network are the ones who have the right to use the licensed frequency spectrum. This means that the secondary network makes advantage of the parent network's unused radio frequency bandwidth. The users and the CR base station make up the system.

Specifically, CR has three features that set it apart from conventional radio:

1. Understands the geographical networks of the applied area
2. This cognitive data allows CR to make intelligent, independent decisions about adjusting its settings on the fly.
3. CR is also able to gain insight from past events and try out alternative setups in novel circumstances.

4.2 COGNITIVE RADIOS AND SPECTRUM SENSING

When compared to traditional radio, the all-intelligent radio network known as "CR" is superior. In contrast to conventional radio, a CR configuration makes full use of all available resources, including the underutilized portion of the radio spectrum. An important benefit of CR is that it can find free spectrum channels and adjust broadcast parameters to make use of many unused frequencies at once. Nonetheless, at this time we urgently want comprehensive surveys and descriptions of the CR-sensing systems. Using CR will also improve the radio's operational behavior. This chapter describes the CR's sensing and interference mechanisms and explains why this arrangement is superior to traditional radios in many ways. CR infrastructure makes use of a wide variety of technologies, including adaptive radio and software defined radio (SDR). For over a century, people have been listening to the radio. Because of the way radio works, anyone using a specific frequency band risks having their broadcasts mixed in with everyone else's. The frequency bands will become extremely congested as the number of users continues to rise. The exponential expansion of wireless technologies over the past few decades has only made matters worse. The specifics of the frequency allocation demonstrate the severe difficulty the expanding demand faces due to the scarcity of newer frequencies. Frequencies are not used in the latest wireless system designs.

It's a challenge for any developing system, as the underlying infrastructure tends to become unsustainable after a few years. Re-allocating the current infrastructure is likewise out of the question. Attempting to solve this enormous problem is what gave rise to the idea of "CR" in the first place. In CR, new users are given access to previously unoccupied channels by making use of existing signal processing capabilities. The new technology was constrained by the need to maintain or improve quality, accommodate an increasing number of users without negatively impacting existing ones, and increase data transfer rates. More efficient use of existing frequency bands through technical means is one such method. To make room for the newer frequencies in the same spectrum, several adjustments need to be made. The hoped-for reforms by those in power have been unsuccessful. Radio, television, satellite, airport control, and so on are all examples of bands. A CR serves multiple purposes throughout its lifetime. The main phases that are considered for the proper functioning of the CR are sensing, analysis, reasoning, and adaptation.

4.3 THE COGNITIVE CYCLE

White-space spectrum detection occurs during the sensing phase. The unoccupied frequency range in space is available for usage.

This unused bandwidth can be utilized for transmissions using CR frequencies. During the sensing stage, we check for signs of activity in both public and private spaces. Second-user interference patterns are something that needs to be checked for throughout the monitoring phase.

Due to the real-time nature of the monitoring, the blanks must be identified with precision. Once the spectrum phase has been sensed, the white space at the optimal frequency for maximum quality utilizing QoS can be located.

There are a few other criteria that the empty area must fulfill before it may be used for transmission. Both primary and secondary users can be transmitted at the same time, although secondary users can switch frequency bands if necessary. To evaluate the quality of the available frequency, common metrics to consider include noise levels, losses, and error rates. It's worth noting that both primary and secondary users have access to the frequency specifications. The CR is built so that after checking the parameters, a suitable frequency can be given for secondary users. In a traditional radio network, spectral efficiency is underutilized. The benefit of CR is that it can use the same frequency for many communications.

When proposing a design for a CR, it is important to consider several capacity restrictions and the often-varied approaches that are used. It is widely agreed that CR has the most efficient use of the spectrum. In terms of

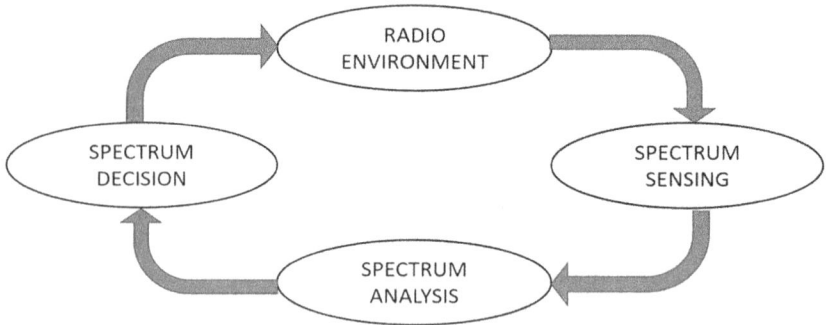

Figure 4.2 CR cycle.

both design and spectrum use, the details regarding the channels, messages, and node shares are essential.

The three most popular methods for handling contextual data in CR are underlay, overlay, and interweave. Each of these three methods has its unique traits. The CR cycle is shown in Figure 4.2.

The requirements for CR include high-quality transmission, compatibility with the existing channels, and the ability to intelligently use the information contained in the frequencies. Overlay, underlay, and interweave are the three forms of channel information utilized by CR.

While implementing the new channels and frequencies, there is a chance of interference. Underlay occurs when the sum of the interference from the new users is less than the average of the frequencies. The quality of the older frequencies is enhanced by the addition of the newer ones in an overlay. The additional frequencies fill in the system's unused spectral gaps in an interwoven pattern. To make room for the new frequencies in the existing frequency range, CR often employs one of these three network technologies.

4.3.1 Underlay

Understanding how new and old frequencies interact is the foundation of CR's initial paradigm. CR is considered a secondary user in this context, alongside conventional radio. When compared to cognitive frequencies, the interference generated by no cognitive frequencies must be smaller than the threshold. For both transmissions to go through, the difference must be smaller than the threshold value. There is more than one approach to taking care of this matter. An alternative is to utilize several antennas to separate mental and physical signals. The second approach involves using a large bandwidth to guarantee the cognitive signals are spread out below the noise level before being beamed to the receiver. These techniques keep interference to a minimum and prevent signals from canceling one other out during transmission. Cognitive signals can have their transmitters optimized such that their output power is kept low enough to keep the threshold constant.

Based on this mechanism, it is reasonable to conclude that the underlay approach can only be employed for localized communications mediated by cognitive signals. The primary cause is the restriction of available resources. Users in many unlicensed bands employ the underlay technique. When evaluating the underlay radio, the interference temperature is the gold standard. This parameter characterizes the antenna's output strength in terms of the radio frequency being transmitted. This strength can be used to determine if the level of interference is excessive. The next diagram illustrates this point. Signal-to-noise ratios could be utilized for calculating the value and level of interference, and the average power received in terms of interference is roughly equivalent to the secondary power constraint at the transmitter. Note that the cognitive signals cannot be delivered if the interference value is at its highest, as this is where the threshold value comes into play. As a result, the maximum power value is kept below the target level.

If you use the underlay approach, your service quality will be equal to the mean of all the transmitter powers. Finding the minimum transmission level follows calculating the maximum transmission power. Priorities for the utilization of the cognitive band may be set differently for various users. With this service, customers can share the same cognitive bands among multiple channels, each of which can send unique data.

4.3.2 Overlay

The main distinction between the overlay and underlay type is that the latter permits the usage of both non-cognitive and cognitive users at the same time. There will be no disruption to either user's transmissions. It's also feasible that there are just two users present, and one of them is keeping their data secret from the other. It's crucial to encode the data such that the non-cognitive user's transmission doesn't mess with the cognitive user's processing. The interference can be prevented if the cognitive user is aware of the non-cognitive user's data transmission type. When information from one channel is available through another, complications occur that could be disastrous for the whole communication network. Any time the two channels are close to one another, this is a real possibility. A communication failure can occur when many problems, such as a delay or a resend, occur at once. To transmit both cognitive and non-cognitive data without any disruption, different encoding methods are used.

4.3.3 Interweave

Knowledge of the information supplied by the non-cognitive channel is necessary for the overlay strategy described above to deliver the data without interference. When it comes to a variety of uses, this channel is crucial. When channels are close together, interference and other problems slow down and reduce the signal quality. On the other side, encoding techniques are necessary for proper message isolation across channels and SNR maintenance during the reception. In this context, interference is the most important metric to consider.

Interweave refers to a set of techniques and strategies used to transfer data in a way that avoids interference and "hops" over any gaps in the network. When the knowledge necessary for non-cognitive transmission is at hand, both cognitive and non-cognitive transmissions can be accomplished. Part of what makes the interweave technique so effective is the emphasis it places on the existence and specifics of transmissions from non-cognitive systems. Complex problems arise when trying to recognize a non-cognitive system as a user when the signal weakens. When information from unlicensed channels and frequencies is added to the preexisting signal band, the system's noise value likewise rises. The fact that the non-cognitive data is not a fixed value presents still another challenge in its detection. It changes through time and evolves. The strength of the signal continues to swing wildly.

Data should be checked repeatedly over a set period as part of the detection technique. User detection can be improved by monitoring unread broadcasts. When doing computations, the false negatives are taken into account as well. Also, because of the signal's inherent volatility, its perceived value may be so low as to be negligible.

4.4 TEMPERATURE INTERFERENCE IN CR

Because of primary and secondary user interference, a crucial method for sensing temperature is implemented. Primary users can be located by listening for their frequency. The noise at the interference can be used as a proxy for the amount of interference that was avoided by reducing the transmitter's output. Limiting the frequency and, by extension, the power of the signal is impractical and could result in signal losses and poor transmission quality, which is why this approach has failed. It's also conceivable for any number of additional interfering noises to materialize. Therefore, the reception measures the radio frequency power before sending it on to the receiver. A receiver's power is quantified by what is known as the interference temperature. In this sort of sense, the signal's energy is the parameter being measured.

This is the simplest approach because it doesn't call for knowledge of the primary signals. The absence of signal data information is a limitation of this approach. Sensing may provide a large number of false positives if the noise value is not accurately calculated. However, the calculation's uncertainty increases if the signal quality is low. Even if multiple signals are present, our approach cannot determine which one is the major one. Sensing the spectrum allows for the determination of numerous transmission parameters and the precise calculation of the available spectrum.

Primarily, this technique ensures that the available frequency gap satisfies the need for efficient transmission of the CR. Feature detection is a subset of SS that includes this type. One such parameter is the spectrum's feature density, which is used to describe a high-quality signal. For the feature detection method to function properly, the design must specify the properties of the principal signal to be detected. Compared to power detection, this method is more involved, but it is more resistant to background noise. Since several other factors are employed to determine the presence of the signal, this method is effective even when the signal strength is low. This approach may also be able to detect several signals and select only the most important ones. For this strategy to work, you need to understand what the primary signals are like. This strategy is effective, but the design process is complex, especially when it comes to characterizing the signals based on properties that are unique to the system being created.

Spectrum analysis is a pivotal stage in the development of CR traits. The detection of white spaces in the spectrum is a crucial step in the process because it allows for the identification of any area of the transmission frequency that is not being used by other users. The other thing that happens during this stage is that the channels already being used by transmitters and users of the spectrum are not changed. At this stage, we analyze the data to make sure there is no overlap between the new empty spots and the ones already in use. It will be up to the secondary users to pick the frequencies that work best for them. The success of the system depends on this step, which is why it is so important. In a nutshell, the findings may be broken down into three stages: dynamic allocation, reasoning, as well as adaptability.

Another technique for checking and distinguishing between primary and secondary signals is called matched filtering, and it does so by looking at the pattern of the waveform of the signal. This technique uses a coherent detection system to separate the primary signal from any interference. To ensure the presence of the primary signal, the detected signal match is compared to the needed signal. Since the signal's specifics are recorded and compared later, this method uses fewer samples and is hence simpler. It is also resistant to noise and poor signal quality. It's a sophisticated design, though, because this approach needs a detailed examination of the signal to be used as a benchmark.

4.5 DYNAMIC ALLOCATION

Allocation is done on a dynamic basis, with the most resources being allocated to the most important users. According to the plan, the secondary users' frequency might be taken by the primary users if the former asks for it. Once secondary users have sensed the available frequencies, they can pick one to use. As a bonus, the layout permits primary and secondary users to share the allocated frequency. This method ensures that all users make efficient use of the available frequency range. Additionally, throughout this period, the principal and secondary users' frequency bands are quickly characterized.

4.6 REASONING

After collecting data, the following step is reasoning to establish the optimal response strategy for frequency allocation. After the band's analysis is complete, the design notifies the secondary users of the allocation details. The sensing and frequency determination techniques are algorithmically diverse. Frequency comparison, selection, and deductive reasoning are all carried out at this point. The algorithm's implementation determines the system's effectiveness. Whether or not a secondary user can be assigned a frequency depends on numerous user factors. This is where all the design smarts come into play. Parameters can also be analyzed with the help of cutting-edge machine-learning techniques. Primary and secondary users shouldn't introduce any noise into the system, as this is the most crucial requirement. However, notwithstanding these restrictions, there should be no unallocated frequencies. Once the optimal values for all the parameters have been determined, frequency allocation can begin.

4.7 ADAPTATION

Following the conclusion of the reasoning phase, the frequencies will be assigned with updated parameters for the new transmission. For CR, this is the final stage of its life cycle. As the CR's cycle completes at this phase, so does its own. Primary and secondary users are currently making use of the available frequencies. The secondary users shouldn't be bothered or experience any interference as a result of the modifications, but the prime users will need to adjust. All necessary time can be spent in this phase of transmission and reception for both primary and secondary users. It would be ideal if interference never occurred, transmissions were error-free, and SNR was kept high. As soon as the transmission is over, the cycle returns to the sensing phase, where the unused frequencies are felt. As long as there are tertiary users who need access to primary transmission frequencies, this process will continue.

4.8 SPECTRUM SENSING IN COGNITIVE RADIOS

One of the most strictly controlled and scarce resources on Earth is the radio spectrum. With the explosion in wireless devices in recent years, there is now a critical shortage of radio spectrum due to its fixed allocation. But numerous static allocation analyses reveal that licensed spectrum bands are underutilized. Spectrum scarcity and underutilization are problems that have been proposed to be addressed by CR. CR relies heavily on SS to identify spectrum gaps. Several SS methods, such as matching filter detection, cyclo-stationary feature detection, and energy detection, have been developed to determine whether or not principal user signals are present. Due to its simplicity, quick sensing time, and cheap computational complexity, energy detection has attracted a lot of attention from researchers. Low signal-to-noise ratios (SNRs) cause a quick decline in performance for traditional detectors because of their sensitivity to noise uncertainty. In this research, we apply Shannon, Tsallis, Kapur, and Renyi entropy-based detection to reduce the impact of noise, and we evaluate their results to determine which method is most effective. The comparison results show that the Renyi entropy is superior to the other entropy approaches. To enhance the efficiency of single-stage detection methods, this research proposes a two-stage SS scheme wherein energy detection serves as the coarse stage and Renyi entropy-based detection serves as the fine stage. Traditional energy detection, entropy-based detection, and the suggested two-stage approaches are compared for performance over an AWGN channel. SS methods can be judged by how well they operate in terms of several different metrics, including the probability of detection, the likelihood of a false alarm, the likelihood of a missed detection, and the shape of the receiver's operating characteristics curve.

Today's wireless communication technologies are constantly developing and improving to meet the ever-evolving needs of society. A lack of available spectrum is being blamed on a lack of flexibility in the way spectrum is allocated, even though this policy has not changed in the face of the growing demand brought on by technological breakthroughs in wireless communications. Rather, recent research on present spectrum allocation reveals that the licensed user at any given area and time is underutilizing the spectrum granted to them. One potential solution to the problems of spectrum scarcity and underutilization is CR.

To make better use of scarce and underutilized frequency bands, CR is an essential technology. SS, spectrum judgment, spectrum sharing/allocation, and spectrum mobility/handoff are the four main tasks/functions of CR. SS is used to locate unoccupied frequencies and detect primary licensees in the area. The purpose of spectrum management is to pick the best available vacancies in the spectrum. To achieve its goals, spectrum sharing must ensure that the gaps between secondary users are shared properly. The goal of spectrum mobility is to keep the lines of communication open while upgrading to more optimal spectrum gaps. Figure 4.3 illustrates the CR's SS.

Among all of CR's roles, SS is widely regarded as the most crucial. Different SS approaches have been proposed over the past few decades, and they can be broken down into two groups: wideband and narrowband,

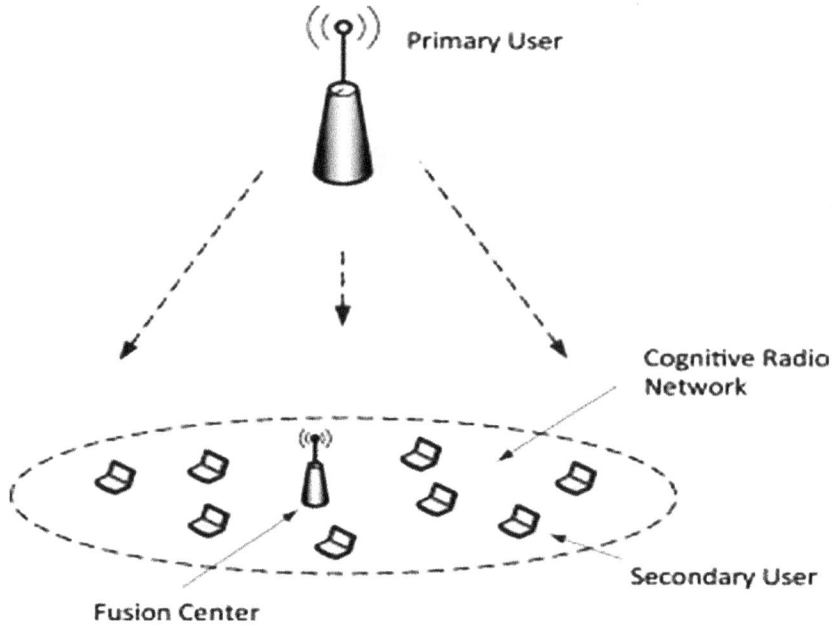

Figure 4.3 Spectrum sensing in CR.

depending on how much bandwidth is needed to adequately sense the spectrum. In contrast to wideband SS, which looks at several frequencies all at once, narrowband SS focuses on just one. Sensing methods can also be broken down into two groups, coherent and non-coherent, depending on whether or not they require any prior knowledge of PU signals. However, there are a few various ways to identify gaps in the spectrum, and they are typically categorized as either transmitter-based, interference-based, or receiver-based. Only transmitter-based detection methods are considered in this analysis.

Energy detecting is often used for SS since it is a noncoherent technique that does not necessitate any prior knowledge of PU signals on the part of the SU receiver. It requires little in the way of processing power and can gather data quickly. However, when the signal-to-noise ratio (SNR) is low, ED performance is drastically diminished due to noise uncertainty. Different approaches have been presented to deal with ED's problems. Due to the noise uncertainty at low SNR values, entropy detection emerges as the most reliable of these techniques. The difficulty of its implementation is similar to that of ED, and it does not necessitate prior knowledge of fundamental signals. A wide variety of entropy detection methods exist, including the entropy of Shannon, Renyi, Kapur, and Tsallis. Therefore, this research investigates the relative merits of these two forms of entropy in the context of creating two-stage SS using standard ED. The primary responsibility of SS in CRNs is to ascertain whether or not a primary user is actively using a specific channel, allowing secondary users to make effective use of the available spectrum. For SS to work, the tried-and-true technique of signal detection must be used. A signal can be detected even in a noisy environment by employing signal detection techniques. As an identification task, signal recognition can be represented as an analytical hypothesis test.

4.9 COGNITIVE RADIOS AND SPECTRUM DATABASE

Every week, TV stations update the FCC database on their use of the RF spectrum for the following week. Instead of using laborious, time-consuming, and costly spectrum scanning techniques, CR equipment can consult this database to find open channels. There is a downside to this approach, though: it is problematic to update the database with dynamic spectrum activity in real time. So, there's a chance that CR gadgets won't be able to take advantage of any available spectrum. A combination strategy helps accommodate the increasing number of RF-using devices. As a result, devices will be able to detect empty frequencies more quickly and precisely, which will boost the quality of service.

As a result of the licensed users' actions and the radio environment, it can be difficult to provide sufficient and dependable spectrum resources for unlicensed users in spectrum database-based CRN. For CRN, we advocate here for an adaptive approach to spectrum access through the use of a spectrum database. When deciding whether or not to use a licensed spectrum, secondary users rely not only on the spectrum information provided by the spectrum database but also on local sensing to verify the actual state of the spectrum. Optimal decision-making models the adaptive sensing and access process by increasing the possible throughput of CRN. Specifically, a dynamic programming technique is created to determine the best sensing and access policy for each SU. The suggested sensing and access policies are shown to be reliable in simulations, allowing for the discovery of spectrum opportunities despite the changing nature of the radio environment.

The use of CR has shown promise as a method to help deal with the spectrum crunch facing the modern wireless communications infrastructure. Secondary (unlicensed) users (SUs) can improve their network's overall performance by taking advantage of spectrum opportunities, which are instances in which they are allowed to use the licensed spectrum band without interfering with the transmissions of primary users. Understanding whether or not the licensed spectrum is available for usage by SUs at a given time is a basic problem in CRN. Several processes that can help a CRN understand its spectral surroundings have been the subject of extensive research. Using the SU's local SS is a straightforward approach. A sensing-transmission cycle governs an SU's actions in this system. In this loop, an SU checks in on the licensed spectrum at regular intervals to look for spectrum openings. Consequently, an SU's sensing capability is crucial, as the performance of CRN is largely dependent on the number of spectrum possibilities that the SUs can uncover.

However, the battery-powered SUs power could quickly run out due to the mandatory SS's periodic identification of the licensed channels. Using a spectrum database is another method to acquire licensed spectrum information while conserving SU power. The geo-location/database techniques for spectrum awareness are being used in the development of the IEEE 802.22 WRAN standard. If the SUs are known to be in a certain area, and a list of licensed transmitters is accessible, then the CRN can utilize this information to figure out which channels are free to use. The Federal Communications Committee has also announced that individual SUs would no longer be permitted to conduct SS. It mandates that SUs make use of database-based spectrum access. To keep track of which frequencies have been claimed by which PUs, a spectrum database is maintained, and the SUs are mandated by the CR base station to consult this information before making their spectrum allocation decisions. But there are restrictions on relying solely on databases to supply details about licensed channels. Spectrum database is unable to real-time information updates and SUs environment awareness. Unfortunately, the local SS approach's crucial sensing overhead limits how quickly it can sense and make judgments. Consequently, there is a cost-benefit analysis involved when deciding between a spectrum database and local sensing inside a CRN.

4.10 COGNITIVE RADIO AND 6G NETWORK

Beginning in 2020, global communication and spectrum sharing will rely significantly on 5G CRN [1]. The backbone of 5G CRN is massive machine-to-machine communication, increased mobile broadband, and low-latency ultrareliable communication. Researchers investigated various essential technologies, including ultra-dense networks, millimeter-wave, and multiple-input multiple-output (MIMO), to accomplish 5G CRN communication.

The fifth generation, on the other hand, will not suffice for 2030 and beyond. Academics and think tanks are interested in 6G CRN communication. Because low latency is critical in 5G, a deterministic network is required to ensure point-to-point inactivity in line with the demanding norms of future mandates. The 6G standard must be more capable of synchronization than the 5G standard. Furthermore, 6G will offer sub-centimeter geographical precision, millisecond position updates, and the ability to cover the greatest amount of ground imaginable [2].

For example, 5G is currently difficult to adopt in rural regions, limiting the utility of some applications such as self-driving cars. To achieve depend-ability, consistency, and efficiency, broadcast networks for ubiquitous services require particular communication satellite and non-terrestrial networks. The communication network is critical for giving a timely reaction in difficult conditions. In 5G, millimeter-wave may provide Gbps data transfer, while Tbps data speed is required in 6G for applications such as 3D high-quality films, VR/AR, and other potential competitor bands such as THz. New access to smart applications will be possible in 6G with the aid of machine learning (ML) and artificial intelligence (AI) due to the usage of exceptional heterogeneous networks, a large number of antennas, a new service-varied communication scenario, and broad bandwidth. Several aspects of a network's operation, such as fault management, security, user experience, services, and energy efficiency, can all benefit from automation. The majority of network traffic in 5G is accounted for by video and audio streaming services. Aside from the previously mentioned purposes and requirements, 5G physical information transmission can shed light on which CRNs are operated by robots. Modern applications of cellular technology include factory logistics and autonomous driving; however, these fields still bring some fresh challenges. In a control system, the central network is overwhelmed when multiple mobile objects transmit control and sensor data. Future technology and demand are propelling research into the use of AI in distributed control systems. AI and similar application networks are projected to dominate CRN traffic demand at 6G, a hitherto uncharted industry that brings significant issues.

To implement 5G CRN, 6G CRN will offer maximum improvements in spectrum efficiency, cost-effectiveness, energy savings, the maximum data rate in Tbps, about ten times reduced latency, nearly 100 times better connection density, and maximum intelligence for automation. Increasing energy efficiency and spectrum efficiency using channel coding methods, current waveforms, multiantenna technologies, and multiple access approaches is difficult without adopting cutting-edge transmission and SS technologies, as well as eliminating air interference. Meanwhile, technologies such as software-defined networks, cognitive service architectures, service-based designs, and cell-free layouts are required to put such a network design into action. However, as we've seen with 5G development, the software has limitations. VRANs that use general-purpose servers rather than dedicated CPUs for a single application can dramatically lower their power consumption. According to current data, 5G networks consume more power than 4G networks at the same bandwidth.

Instead, we must build a network that is interesting from the start while not surpassing the power constraints of the previous generation. As a result, a new computing paradigm is necessary for 6G if we are to provide all of the benefits of notarization without increasing energy expenditures. Three major computing technologies that contribute to distributed computing, low latency networks, and synchronization are edge, fog, and cloud computing. To compensate for the disadvantages of 5G short packet, low latency service with high data rate, reliable delivery, the Internet of things, and system coverage are all required. 6G CRN must improve data rate or focus on people if it is to fulfill the needs of CRN communication in 2030 and after [3,4]. The availability of such resources will need a paradigm shift in 6G CRN communication.

The minimal standards for all CRN communications are as follows. Energy conservation, AI-enhanced communication, and safer data storage are all possibilities. In the flesh, cyberspace, data transport rates that are extremely high, excellent bandwidth, machine-to-machine communication, congestion in the backhaul, and access networks has been reduced. The simulation in 6G CRN is predicted to be 1,000 times more powerful than in 5G. URLLC, a key component of 5G, will also be critical in 6G communication, ensuring an E2E delay of less than 1 millisecond. When compared to 5G, 6G will offer higher spectral efficiency in volume. Longer battery life will be possible with 6G technology. Mobile devices will not require their charging systems under 6G technology.

The following sections look at some of the most critical technologies that will power the 6G CR-Network.

- AI is a novel and important component of 6G CR communication [5]. AI was not used in the 4G CRN infrastructure. Meanwhile, 5G is providing some assistance to AI technology. Despite this, 6G is planned to enable full AI support for automated services. Thanks to advances in ML, a more perceptive CRN for 6G real-time communication is on the horizon. When AI is used in CR communication, it streamlines operations and improves data in real time. With the assistance of many analyses, AI can develop a complex target path.
- To ensure that all expected devices may connect to the network and that the network can connect to the networks, optical radio technology (ORT) is proposed for 6G CR communication [6]. Light fidelity telepathy, FSO telepathy, optical camera telepathy, and visible light telepathy are all examples of ORT. ORT transmission data rates and latency are incredibly low and secure.
- One such method is the use of MIMO technology [7]. Enhanced MIMO approaches improve spectrum efficiency. Because of this, 6G communication networks will benefit from massive MIMO technology.
- To fulfill big data ambitions, future communication system networks will need to contain blockchain technology. Blockchains, in and of themselves, are merely one type of distributed ledger system. Distributed ledgers, on the other hand, are defined as databases dispersed over numerous computers. Each node copies the ledger and securely stores it. Peer-to-peer networks round out the blockchain. It is even conceivable without the involvement of a centralized body. In blocks, the blockchain collects, sorts, and arranges information. The bits are consistent with one another and can be stitched together using cryptanalysis. The blockchain enhances security, interoperability,

reliability, privacy, and scalability, to name a few benefits [8]. As a result, the 6G CR communication system will be able to provide a plethora of services, such as tracking massive amounts of data, sharing information between devices, facilitating autonomous inter-actions between various Internet of Things systems, and maintaining a consistently high level of connectivity.

Privacy safeguards are a must-have for critical infrastructure networks in the 6G era. Data loss, theft, and even physical assaults are all real risks in CRNs. Due to the lack of apparent collaboration between primary and secondary user communication in CRNs, these networks are more vulner-able to security threats than conventional wireless networks. Because the CRN ecosystem relies on sensing information, certain security problems created by CR can alter node behavior. Since learning is central to CR, a faulty assumption about the surrounding environment can lead to a bad call. As a result, hostile attacks will use this vulnerability to cause permanent changes in the target behavior. Unauthorized parties can delete, sniff, or alter CR data. Security in CRNs must be sufficiently resilient to withstand attacks and threats. Additional security concerns in 6G CRNs include the following: tampering with data, unauthorized access to improper interruption of the primary user, private data, disabling the common control channel through an artificial yielding blockage issue, disabling the idle channel for unlicensed users, and injecting false data. 6G CRN benefits from 5G technology's reduced latency, efficient transmission, high reliability, and secure services. The greater the likelihood that they will carry out protecting their clients at the expense of compliance with regulations.

4.11 CONCLUSION

The nascent technology of CR can be used in a variety of contexts. A few studies have been conducted in this area, making surveys all the more important. This overview explains how CR, networks, and sensors all work on the most fundamental level.

We now have a good understanding of the most crucial aspects of CR and how they work. The next phase entails the creation of a CR. Phases of sensing, analyzing, reasoning, and adapting are explored to round out the full cognitive life cycle. Researchers in this area will benefit from the information offered in this publication. The results also demonstrate several scenarios and procedures for efficiently utilizing CR via various systems for sensing and transmission. The future issues of the 5G system can be met by the incorporation of CR. This technology supports the rising volume of mobile data traffic while delivering a wide variety of useful services. Small cells coexist with micro and macro cells in networks, and efficient spectrum management, especially spectrum sharing, leads to increased performance. Future wireless networks will also benefit from the implementation of advanced wireless communication technologies such as three-dimensional (3D) beamforming, massive MIMO, millimeter wave communication, and visible light communications. Antenna requirements for enhanced 5G performance in CR are also covered here. An adaptive sensing and access approach for spectrum database-driven CRN is developed and evaluated in this book chapter. The proposed technique seeks to decide whether to directly use the information from the spectrum database or to apply the local sensing to obtain the channel information based on the dynamic channel conditions. Taking into account the SUs' access operations, we develop a throughput maximization problem to determine not only the best time to do local sensing but also the best sensing method to use. Comparing the suggested sensing and access with two existing mechanisms, the illustrative results show that the proposed method achieves significantly higher throughput and lesser interference to PU's transmission.

REFERENCES

1. Venkatesan, G. A., Kulkarni, A. V. , Menon, D, & Swetha, N. P. (2019). *Role of Cognitive Radio in 5G*. Helix. 10.29042/2019-4850-4854
2. Srivastava, A., Gupta, M. S., & Kaur, G. (2020). Energy efficient transmission trends towards future green cognitive radio networks (5G): Progress, taxonomy and open challenges. *Journal of Network and Computer Applications*, 168, 102760. 10.1016/j.jnca.2020.102760
3. Zong, B., Fan, C., Wang, X., Duan, X., Wang, B., & Wang, J. (2019). 6G technologies: Key drivers, core requirements, system architectures, and enabling technologies. *IEEE Vehicular Technology Magazine*, 14(3), 18–27. 10.1109/mvt.2019.2921398
4. Dang, S., Amin, O., Shihada, B., & Alouini, M. (2020). What should 6G be? *Nature Electronics*, 3(1), 20–29. 10.1038/s41928-019-0355-6
5. Stoica, R., & Freitas de Abreu, G. T. (2019). Frame-theoretic precoding and beamforming design for robust mm wave channel estimation. 2019 IEEE Wireless Communications and Networking Conference (WCNC). 10.1109/wcnc.2019.8885599
6. Alsharif, M. H., Kelechi, A. H., Albreem, M. A., Chaudhry, S. A., Zia, M. S., & Kim, S. (2020). Sixth generation (6G) wireless networks: Vision, research activities, challenges and potential solutions. *Symmetry*, 12(4), 676. 10.3390/sym12040676
7. Attarifar, M., Abbasfar, A., & Lozano, A. (2019). Modified conjugate beamforming for cell-free massive MIMO. *IEEE Wireless Communications Letters*, 8(2), 616–619. 10.1109/lwc.2018.2890470
8. Taherdoost, H. (2022). The role of smart contract blockchain in 6G wireless communication system. *Procedia Computer Science*, 215, 44–50. 10.1016/j.procs.2022.12.005

Chapter 5

A novel energy-efficient optimization technique for intelligent transportation systems

Shaik Rajak[1], Inbarasan Muniraj[2],
Poongundran Selvaprabhu[3], Vetriveeran Rajamani[2], and
Sunil Chinnadurai[1]

[1]Department of Communication Engineering, School of Engineering and
Sciences (SEAS), Andhra Pradesh, India
[2]Deptartment of ECE, Alliance University, Bangalore, Karnataka, India
[3]Department of ECE, Vellore Institute of Technology, Vellore, Tamil Nadu,
India

5.1 INTRODUCTION

5.1.1 History

Primitive metasurfaces were traditionally employed to deliver a set of electromagnetic responses for certain incident radiation (frequency, incident orientation, and polarization) that was presumptively determined. Configurable metasurfaces were created as a result of the rapid realization that the capacity to flexibly tune the features of the metasurface, either to adjust the output signal capabilities or to adjust it to an input signal with distinct features, would greatly increase the potential of metasurfaces for real world applications. Due to their capacity to control electromagnetic waves, metasurfaces, the ultraportable, 2D form of metamaterials, have notably gained a lot of attention. Reconfigurable and programming metamaterials have seen their extent and impact on real-world applications substantially expanded by recent advancements. These useful blocks of materials may behave as electromagnetic field-shaping artificial blocks and have a significant influence on imaging, connectivity, and sensing applications. In order to get the necessary reaction to electromagnetic waves, the building blocks can be composed of a variety of material properties (dielectric, metallic, and semiconducting). Metamaterials can be identified by effective, normalized surface component properties like electric and magnetic layer conductivities or impedances because of their subwavelength regularity. These reconfigurable blocks can offer remarkable traits and characteristics for controlling electromagnetic waves, such as precise absorption, altered reflection, waveform directing, polarization control, and dispersal design, relying on the resonant characteristics and the form of

DOI: 10.1201/9781003369028-5

the meta-atoms. These metamaterials have been identified by different names as Reconfigurable Intelligent Surfaces (RIS), Intelligent Reflecting Surfaces (IRS), Maeramaterails (MM), and Metasurfaces (MS).

5.1.2 Performance of wireless networks by using IRS

In recognition of the recent increase in transportation demand, intelligent transportation systems (ITS) were implemented to provide speedy, safe, and comfortable modes of travel facility (Dimitrakopoulos et al., 2010). ITS uses a cutting-edge wireless communication system that combines intelligent, seamless connections, techniques, and other advancements to make it easier for people to use roads and cars (Zhang et al., 2011). In order to improve the performance of ITS, some research merged vehicle-to-vehicle (V2V) and vehicle-to-infrastructure (V2I) networks in vehicular technology (Dey et al., 2016). By integrating V2V and V2I networks leveraging most source distribution as well as spectrum sharing, future ITS will be more effective and efficient. However, with the rapid advancement of automated cars, it is getting harder to meet ITS requirements (Smith et al., 2005). A significant amount of data has to be delivered very quickly in ITS by using 6G communications, which need more transmit power. As a result, ITS requires precise and practically latency-free connectivity for more efficient and secure operations. High buildings and other barriers may potentially result in a signal interruption in an ITS system. The most revolutionary technology that can enhance the network capacity of future wireless communication systems is thought to be intelligent reflecting surfaces. Therefore, new research on wireless communication networks established the metasurfaces with reflecting components to create the reconfigure-based environment between the sender and receiver. To improve the functionality of the communication network, the phase shifts of the reflection components in IRS have been modified to rearrange the transmission signal between the sender and the receiver.

In Özdogan et al. (2020), physical optics methods are employed to calculate the pathloss factor for an IRS, where the obtained beam radio band is inversely proportional to the number of IRS elements, to regulate and reflect the received signals from a distant field. As the number of IRS elements increases, the RF chain lengthens and the design becomes more complex. The MISO network has been investigated for beamforming techniques using IRS components. The system with enormous components requires several RF chains, which made the CSI estimation more complex at IRS. The particle swam optimization technique was used at BS and IRS without CSI to provide close to optimum results for beamforming (Souto et al., 2020). When least transmit power results from the non-orthogonal multiple access (NOMA) and orthogonal-multiple-access (OMA) models are examined, it is discovered that the NOMA technique has not fully demonstrated its superiority to OMA without IRS. Therefore, a Low complexity method to find the nearly optimal results with an IRS-assisted NOMA technique has been suggested.

To investigate how hardware problems in the IRS-aided NOMA method affect performance, spatial division multiple access (SDMA) is used (Ding et al., 2020). To determine the up-link and downlink transmission effectiveness for the user at a far distance from the BS with the Nakagami-m fading conditions, IRS-aided NOMA, and OMA are taken into account. The IRS-aided wireless network outperformed the full-duplex decode-and-forward (DF) relay, according to the results (Cheng et al., 2021). Shaikh et al. (2021) studied the SE, EE, and outage probabilities and evaluated how transceiver limitations impact IRS- assisted wireless systems. They discovered that after a given SNR, SE decreases even with an increase in the IRS elements and transmission power by using the Gaussian distortion model at the user. The projected gradient method (PGM) was used in Perović et al. (2021) to resolve the joint optimization of the transmission power and IRS elements for the MIMO system. The block coordinate descent (BCD) approach is used to simplify the non-convex joint objective functions for the IRS-aided multi-cell MIMO system. Later, majorization minimization (MM) and complex circle manifold (CCM) are used to calculate phase-shift tuning in order to improve cell-edge user performance (Pan et al., 2020). Furthermore, to reflect the incident signal toward the user without the deployment of a relay, the IRS-aided framework is built with a significant number of passive devices that work at millimeter and sub-millimeter frequency ranges (Di Renzo et al., 2020). For high-frequency IRS mm-wave systems, Wang et al. (2021) utilized the super resolution (SR) network depending on the Least Square (LS) estimate method to simplify channel estimation. By carefully choosing the amplitude reflecting coefficient and implementing the optimal scheduling method, IRS can reduce energy usage while enhancing the connectivity of the Internet of Things (IoT) beyond the capabilities of 5G cellular networks (Yu et al., 2020). The IRS-based system's actual power utilization in a realistic outdoor setting has been shown in order to create an energy-efficient framework.

To determine the IRS phase parameters and allocate transmit power, the sequential fractional programming method was used, which outperforms multiple antenna AF relaying (Huang et al., 2019). Numerous previous publications specifically examined the effectiveness of wireless networks with IRS assistance in terms of data transmission or maximizing EE. In order to address IRS reflection design, channel capacity, and resource scheduling optimization, the system architecture with simultaneous wireless information and power transfer (SWIPT) was investigated in Wu et al. (2022). IRS-based jammers that utilize BCD, semidefinite relaxation, and the Gaussian randomization method significantly lower the signal power of eavesdroppers (Lyu et al., 2020).

The IRS block offered improved system results in comparison to the enormous MIMO due to its vast area when compared to individual multiple antenna terminals (Hu et al., 2018). However, according to an analysis of the power scaling rule, the SNR of the one-user IRS cannot compete with the SNR value of a large MIMO. In order to solve this issue, the IRS frame should contain many reflecting components (Björnson et al., 2019). The large-sized IRS block deployment will make wireless design architecture and calculation of signal propagation in varied situations more challenging.

Deep learning techniques are employed for IRS-aided wireless systems to address the aforementioned issue (Taha et al., 2021). A full-duplex relay connecting two IRS blocks was presented in (Alrabeiah et al., 2020), which then related it to AF and DF relays. Results from these authors' research have demonstrated that hybrid IRS-relay-aided networks performed better than regular relays and IRS networks (Abdullah et al., 2021). Later in Björnson et al. (2020), the authors evaluated the achievable rate and EE of single-input-single output (SISO) communications networks, illuminating the importance of IRS elements. The IRS-aided vehicular wireless frame-work BS with IRS and ground-based vehicles to improve QoS and the model under consideration aimed to maximize the V2I data rate while maintaining V2V SINR with a variety of channel conditions. By using the double-phased alternating optimization technique, the aforementioned optimization is made simpler. According to simulations, the IRS can reduce channel losses and increase V2I sum capability in automotive networks. The majority of the previous works mainly discussed how the RIS is used in a particular setting, such as fixed user positions. Concentrated on the modern smart city, moreover, mobile edge computing (MEC) and other cutting-edge solutions are being suggested to cope with the massive data congestion (Duan et al., 2020; Duan et al., 2022).

In addition, Song et al. (2022) elaborated on the effectiveness of the IRS for the ITS and also discussed the DL techniques to improve the perform-ance of DL-enabled IRS for the ITS. The hybrid relay-IRS-supported wireless network has examined the EE with various IRS elements versus SNR as well as with multiple IRS blocks instead of relays (Rajak et al., 2022). The usage of IRS in high-frequency ranges wireless networks for vehicles has been intended to optimize the distribution of resources for autonomous vehicles. Recently, contemporary public transit has IRS involvement in smart transportation has been explored to strengthen the signals while using less transmission power (Zhu et al., 2021; Lu, 2021). However, as aforementioned research works mostly focused on maximizing the EE with perfect or partial CSI, whereas in this chapter we obtained the EE by reducing the channel estimation complexity by using clustering the IRS elements and determined the EE with multiple-IRS blocks for the application of ITS as illustrated in Figure 5.1.

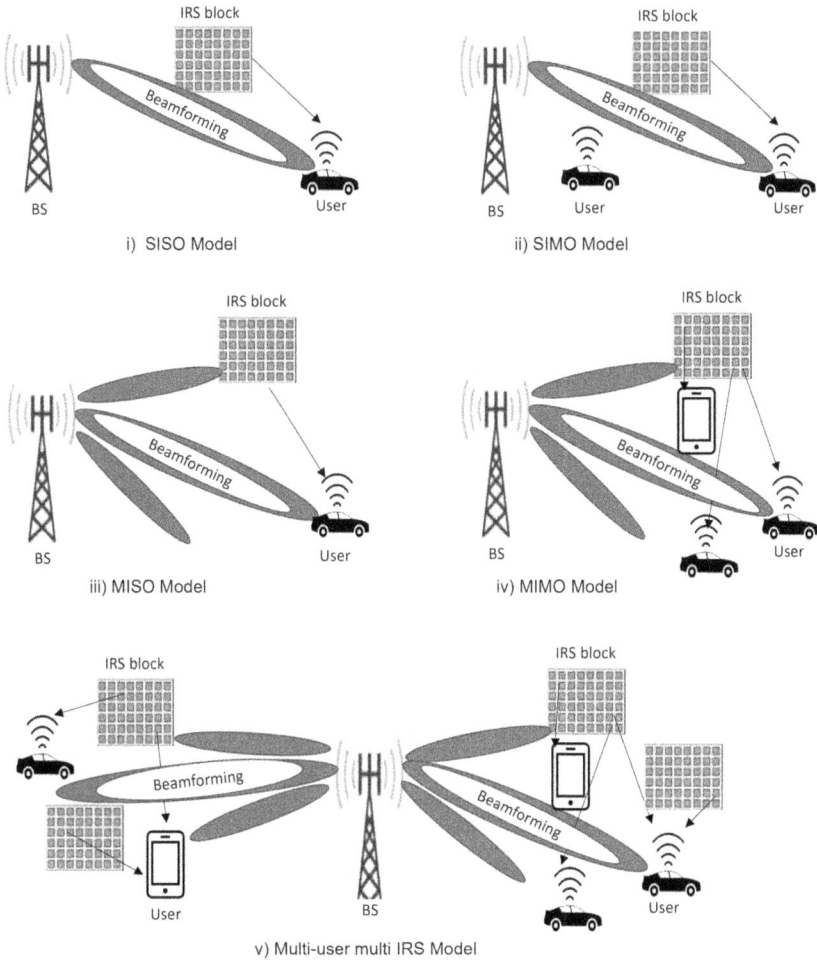

i) SISO Model

ii) SIMO Model

iii) MISO Model

iv) MIMO Model

v) Multi-user multi IRS Model

Figure 5.1 IRS for various wireless communication systems.

5.1.3 Role of IRS in healthcare

In the present scenario healthcare system has been connected with communication networks to provide more precise and fast services. This approach requires robust and continuous connectivity for healthcare devices. As the number of devices increases, they need more data rate, and also all the devices should need seamless data transfer. So, the IRS blocks can be implemented to fulfill the requirements of healthcare systems. As shown in Figure 5.2, the insertion of IRS blocks in the hospital provides continuous communication among all the devices and helps the healthcare system to monitor the patient's condition as well report to the doctors.

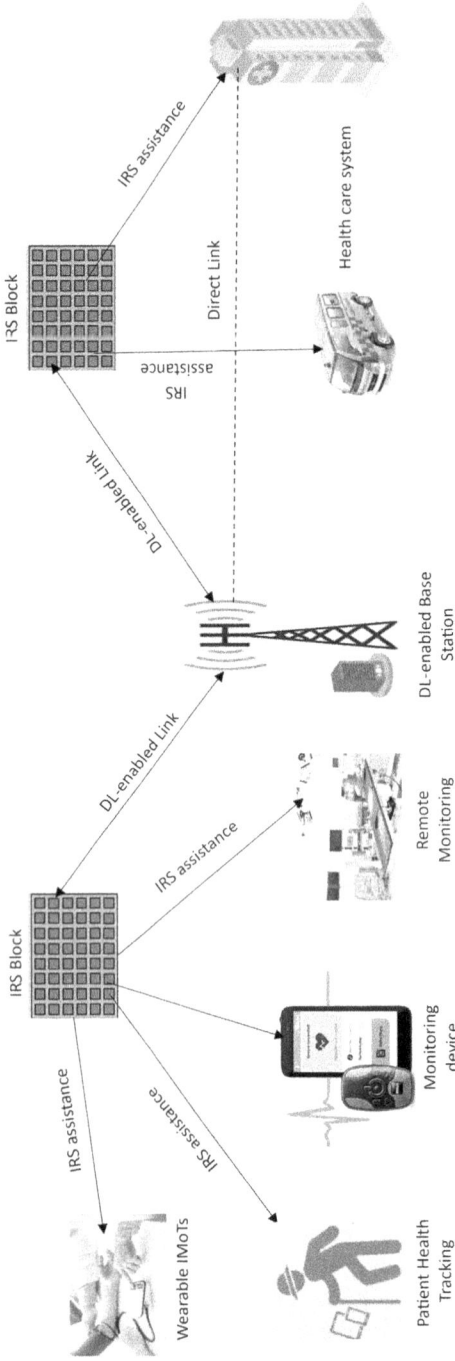

Figure 5.2 IRS for various wireless communication systems.

5.2 ENERGY EFFICIENCY OPTIMIZATION WITH IRS

5.2.1 Conventional methods to optimize the EE

With several individuals, succeeding wireless systems and ITS require high throughput, which utilizes more power. Additionally, as ITS demands faster connectivity with almost zero delays as well as robust and secure communication, IRS can assist ITS in providing a significant amount of data while using less power than traditional relays. The EE of the network is optimized by the IRS, which reflects the incoming signal toward the users without requiring any additional transmission power.

The IRS-aided wireless network is highlighted as an excellent approach to enhancing. In (Wu et al., 2019) IRS, three-dimensional broadcasting was accomplished without the need for RF chains by using wireless communication technologies that reflect the signal with a programmable phase shift. To reduce transmission power, active beamforming at the BS and passive beamforming at the IRS have been merged. In order to boost the SNR while using the least amount of transmitted power, semidefinite relaxation and alternating optimization techniques are used.

A downlink wireless MISO system with diversified IRS that can vary the on-off state according to network specifications has been created in Yang et al. (2022). IRS phase shifts, transmit beamforming, and IRS on-off status was treated as a single optimization problem with the fewest restrictions in order to enhance the EE. Phase optimization and transmission beamforming were resolved for single and multiple-user cases utilizing the SCA approaches. The IRS on/off difficulty was simplified with the dual technique for single-user settings, in which a parametric approach based on the Dinkelbach process was utilized to refine the fractional form. In contrast, the greedy technique is used to address the IRS on-off problem in situations with several users.

The combined optimization of the channel estimation matrices at the BS and reflecting elements of the IRS was explored to enhance the weighted minimum rate (WMR) across all users under the transmission power and unit modulus limits (Peng et al., 2021). The integrated optimization model was more challenging to resolve using standard techniques. Therefore, employing the weighted minimal mean square error and adding certain additional components, the provided BCD approaches correctly convert and resolved the original statement. The computing cost of the BCD algorithm was further reduced via minorization-maximization techniques.

5.2.2 Optimization techniques

The optimization model is challenging to solve in two ways. 1) The unit modulus limitation or the controllable phase shifts make the optimization problem non-convex/NP-hard; 2) The phase shifts and the beamforming vectors are interrelated. As a result, finding a globally ideal outcome is challenging. Nevertheless, the majority of the research done so far has focused on locating computationally simple, highly effective local solutions. Moreover, several studies have been demonstrated that suboptimal approaches can result in higher system stability compared to systems without RISs. The optimization issue transforms into a typical beamforming design issue, which has been thoroughly researched in the literature if the phase shifts are provided. In order to separate the optimization parameters, alternating optimization (AO) techniques are typically used. Advanced techniques like the weighted minimal mean-square error (WMMSE) method or fractional programming (FP) are frequently used to transform the initial unsolvable issue into an approximation problem.

5.2.3 Relaxation and projection

Relaxation and projecting: It is possible to rewrite the unit modulus restriction on the phase shift. Using this method, the non-convex limitation is first relaxed to the convex restriction, after which the result is projected onto the unit-modulus constraint.

5.2.4 Majorization-minimization (MM)

Another popular method for enhancing a RIS's phase shift is the MM method. The MM method is an iterative optimization algorithm that breaks down a challenging issue into a number of easier-to-solve smaller problems.

5.2.5 DL/ML-based techniques for IRS-aided networks

The IRS-supported vehicular wireless network BS using IRS and ground-based vehicles to improve QoS was explored in Chen et al. (2021). The concept under consideration aimed to maximize the V2I data rate while maintaining V2V SINR with a variety of channel links. By using the dual-phase alternating optimization technique, the aforementioned optimization is greatly simplified. The combined power allocation and IRS radiation efficiency were addressed in the initial phase, and phase two observed the implementation of the spectrum utilization procedure for both the V2I and V2V. According to simulations, the IRS can reduce channel losses and increase V2I sum capability in automotive networks. Analyses of safety for automated vehicle environments heavily rely on the accurate identification of driving behavior and prediction of traffic flows.

In Dampahalage et al. (2020), a substantial number of components at the IRS were used to describe how the phase dispersion of an IRS-assisted MIMO network has been operated. The phase of the reflective components might be adjusted due to DL-based methods. Two deep neural networks (DNN) are positioned at the pilot signals that were obtained from the distributed signals. First, the outcomes are contrasted using DNN with the conventional least-square (LS) estimator-based method. The secondary DNN is designed to detect the lowest pilot pattern and to beamform and reflect the appropriate phases.

In several studies, it has been demonstrated that DL approaches perform better for channel estimation in automotive networks. The research works in Chen et al. (2019) gathered real-world information with appropriate IQ values in a range of outdoor settings and matched it to the observed data set of the IEEE 802.11 P model. According to simulated data, the deep-learning-based channel estimation surpasses the traditional auto-regression architecture in vehicular communications and offers more precise communication linkages.

5.3 SYSTEM MODEL AND PROBLEM DESIGN

5.3.1 Channel estimation

Without the use of active relay nodes, IRS is a passive elements block that controls the amplitude and/or phase of the received wave utilizing a large number of passive reflecting components. Thus, IRS can considerably save hardware costs and energy use. These facilities call for an in-depth familiarity with the signals and transmission channels. Two key obstacles to channel estimates are the passive nature of IRS-assisted transmissions and the vast quantity of IRS components, which cannot be eliminated by conventional signal processing techniques. To estimate the channel conditions in various situations, we must create new frameworks for channel estimation.

A three-phased architecture was established by IRS-assisted multi-user communications networks to assess the channel state. When IRS is turned off, phase-I considers the user-BS direct channel, as well as the user-IRS-BS, reflected channels before estimating the channel between both the user and BS. When Phase-II IRS is activated, only one user is permitted to broadcast the pilot symbols needed to determine the channels that the IRS has reflected. By allocating weight to reflect components, the third step calculates the channel environment under which other users can broadcast pilot symbols and their IRS-reflected pathways. Examining both the ideal and real-world instances with and without noise at the receiver. By adding more antennae, the optimum case pilot sequence has been reduced to a minimum. In the real-world scenario with noise, linear minimum mean squared error (LMMSE) was helpful in estimating the channel information for each of the three phases (Wang et al., 2020).

In order to estimate the channel state information, the IRS-enabled inter-vehicle system used Fox's H function decentralized structure as opposed to

Rayleigh/Nakagami/Rician spread. The SNR of the receiver end, route loss, and efficiency were then determined using a dependent mixture of Gaussian (MoG) dispersion, which was confirmed using Monte-Carlo simulations, in place of the well-known central limit theorem. Additionally, numerical numbers illustrated the MoG's strategies for producing successful performance with a small or large number of IRS reflecting components (Kong et al., 2021).

Wireless communications systems supported by the IRS make it more difficult to get accurate channel status data. Many reflecting components lack the capacity to send and receive signals. So, using the strongest channel impulse response maximization (SCM) approaches, Zheng et al. (2020) proposed a workable transmission system for the IRS-aide OFDM system where IRS components are supposed to be in the ON position at all times.

5.3.2 Design and analysis of EE

In this section, we represent the system architecture while considering the single BS with M antennas and ITS supported by IRS with N passive reflection elements as shown in Figure 5.3. The Channel between the IRS to BS, ITS to IRS, and the direct path from ITS to BS are denoted as $H_I \in C^{MXN}$, $h_C \in C^{NX1}$, $h_D \in C^{MX1}$ respectively. The transmitted signals may experience phase variations as a result of the reflective nature of IRS components. Such phase changes are regarded as one of Q distinct levels. For the sake of convenience, we suppose that these phase shifts choose one

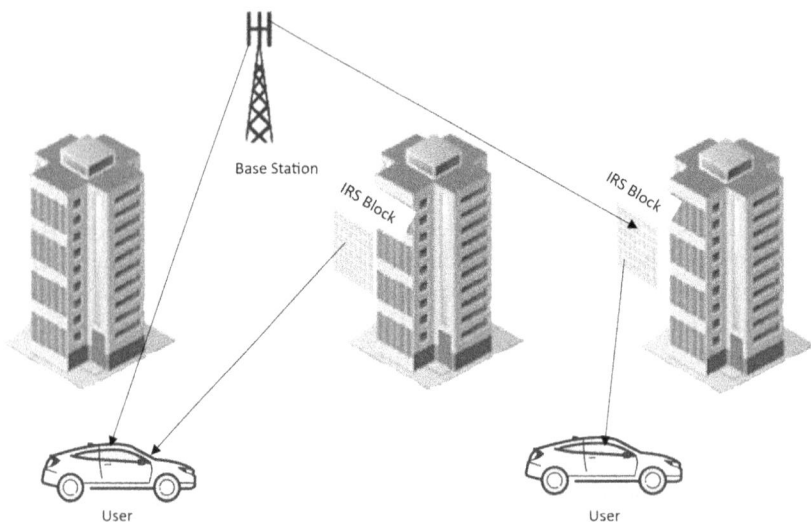

Figure 5.3 System model.

of the values that result from equally utilizing the range $[0, 2\pi]$. Hence, every reflected element's distinct collection of phase shifts is determined by

$$S = (0, \quad \phi 1, \phi 2, .., (Q - 1)\phi) \tag{5.1}$$

Where ϕ is the $2\pi/K$ and the jth IRS reflection phase shift belongs to $\phi \in S$. We defined the reflection matrix for the IRS as $\Theta = \text{diag}\ (e^{j\phi 1}, e^{j\phi 2}, e^{j\phi 3}, ..., e^{j\phi N})$. We derived the message signal m received at the BS, which is transmitted from the ITS as,

$$y = (h_D + H_I\ \Theta\ h_C)m + z, \tag{5.2}$$

where z is the complex additive white Gaussian noise (AWGN) with CN $(0, \sigma 2)$ present at the BS. To decode the message signal m the BS adopts the beamforming vector W, as,

$$\tilde{y} = W^H(h_D + H_I\ \Theta\ h_C)m + W^H z. \tag{5.3}$$

We considered maximal ratio combining (MRC) at the BS, where $W = (h_D + H_I \Theta h_c)$. The signal-to-noise ratio with transmit power P is expressed as,

$$SNR = \frac{P\|h_D + H_I \Theta h_c\|^2}{\sigma^2} \tag{5.4}$$

From the eqn (4), we can determine the achievable rate as,

$$R = \log_2(1 + SN\ R)\ \text{bit/s/Hz}. \tag{5.5}$$

We denote P_T is the total power consumed at IRS and BS, which can be derived as

$$P_T = P_{BS} + P_c + N\ P_e, \tag{5.6}$$

where P_{BS}, P_c, and P_e denote the power consumed by the BS, ITS vehicle c, and each element of IRS respectively. And the N represents the number of IRS elements. From (5) and (6), the EE is calculated by using the expression

$$EE = \frac{\text{Rate}(R)}{\text{Total power}\ (P_T)}. \tag{5.7}$$

From the above equations (5.5), and (5.7) we confirm that EE for the maximum achievable rate mostly depends on the phase shift optimization. To attain optimal EE, we need to tune the IRS elements and find the optimized phase shift matrix and find the channel state information among all the components.

For the implementation of passive beamforming, every channel of the proposed IRS-aided system needs to be estimated. There are unique channels over every reflecting component of the IRS, in addition to the direct link between the ITS and the BS. Typically, an IRS comprises of a lot of reflecting components, as a result, channel estimation becomes more difficult. We suggest two-phase optimization strategies in the following subtopic that can be used to minimize the channel estimation complexity and determine the EE.

5.3.3 IRS phase optimization by using clustering

The phase optimization is done by dividing the entire IRS block into various small numbers of clusters to reduce the channel estimation complexity. In this case, the entire cluster of components is seen as a component. Therefore, rather than having to estimate the channel for every group of reflecting components, we simply need to do it for each cluster. Additionally, instead of proceeding over each separate element, the phase optimization technique can be conducted while taking into account clusters.

Figure 5.4 depicts a 9×9 reflecting block. It is separated into small sub-clusters of size 3. Every sub-cluster is taken into account as a separate

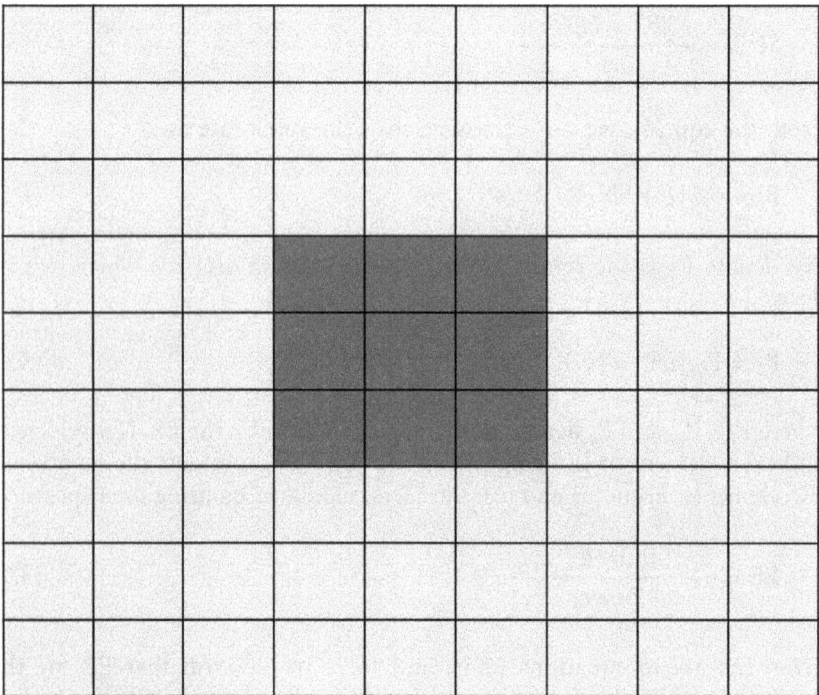

Figure 5.4 IRS block divided into 3X3 sub-cluster.

reflecting component. As a result, a 3×3 reflecting block may successfully perform passive beamforming. This minimizes the difficulty of the successive refinement algorithm (Wu et al., 2020) implementation and the overhead associated with channel estimation. Following the discovery of the phase shifts, all of the reflection elements in the IRS block have their phase shifts modified to identical values for the sub-cluster.

5.3.4 Passive beamforming for the end user based on location

The signal incident at the IRS block can be reflected and re-scattered the same into all directions. The path losses of separate links across the IRS are combined to provide an efficient path loss. As a result, there will often be a greater direct relationship between the ITS and BS. But when the ITS to BS connection is poor and the communication over IRS has robust line-of-sight connectivity, the IRS-assisted network has more advantages compared to a conventional system. Passive beamforming based on end-user location will be possible in this case since line-of-sight connectivity will be crucial. Here, the network will measure the location of the objects as well as their directions of travel and arrival. In this scenario, the channel matrix can be recreated by using the well-known antenna positions.

The LoS channels from BS to IRS or IRS to BS can be derived by considering the planar array at the BS and IRS as below,

$$h_{IRS,\,LOS} = \sqrt{L_{Los}} \, \exp\left(\frac{-j2\pi l}{\lambda} a_{BS}(\Phi^{BS}, \theta^{BS}) a_{IRS}^{h}(\Phi^{IRS}, \theta^{IRS})\right) \qquad (5.8)$$

where BS and IRS are separated by the distance l, the array response of azimuth and elevation angles at the BS are denoted as $a_{BS}(\Phi^{BS}, \theta^{BS}) \in C^{MX1}$ as well as array response at the IRS with azimuth and elevation angles represented by $a_{IRS}^{h}(\Phi^{IRS}, \theta^{IRS}) \in C^{NX1}$. On the basis of the projected LoS channel, the system conducts passive beamforming. Only the arrival and departure angles for the entire reflection matrix need to be estimated with the aforementioned technique. Using the known CSI dimensions of the antenna, the channel matrix could be recreated. In contrast to channel estimation for separate reflecting components, this requires less complexity.

5.4 NUMERICAL RESULTS

This section provides the outcomes from numerical simulations that we conducted to verify the algorithm and illustrate more about the IRS-assisted

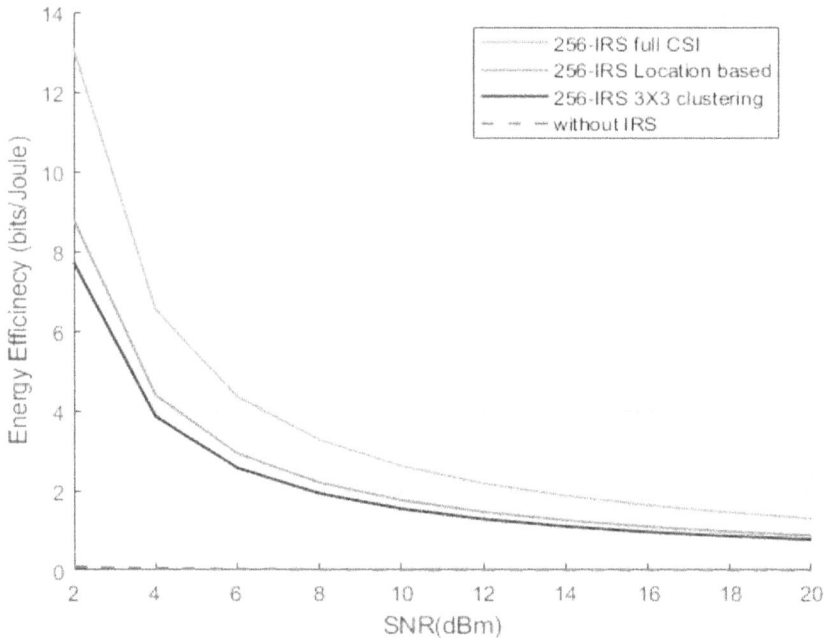

Figure 5.5 EE vs SNR.

ITS. For the simulations in this work, we used millimeter wave frequency ranges with a carrier frequency of 24.5 GHz. A Rician fading channel has been employed to depict each of the relevant channels. The 3GPP TR 38.901 UMi Street Canyon path loss model is utilized to estimate the path loss.

Figure 5.5 describes the EE versus SNR of IRS-assisted ITS with the range of SNR values. Initially, the EE is high and it started decreasing gradually for both the proposed clustering and location-based methods. However, a location-based method shows better performance compared to without IRS and a clustering-based technique. As we can clearly observe that the EE for location-based at SNR 6 dBm is more than 4 bits/Joule whereas for clustering based the EE is 2.5 bits/Joule. So, from the above observations, it is noticed that location-based IRS-aided ITS is more efficient than the clustering technique.

In Figure 5.6, we determined the EE versus SNR for single and double IRS-aided ITS. The EE of the single IRS block-aided system has lower values than double IRS blocks for both the location and clustering-based techniques. In particular, double IRS location-based systems EE at the SNR = 15 dBm is 14 bits/Joule at the same time for double IRS clustering models EE is 10 bits/Joule. From the above results, it is clearly confirmed that the double IRS location-based ITS outperforms the clustering-based system for all SNR values.

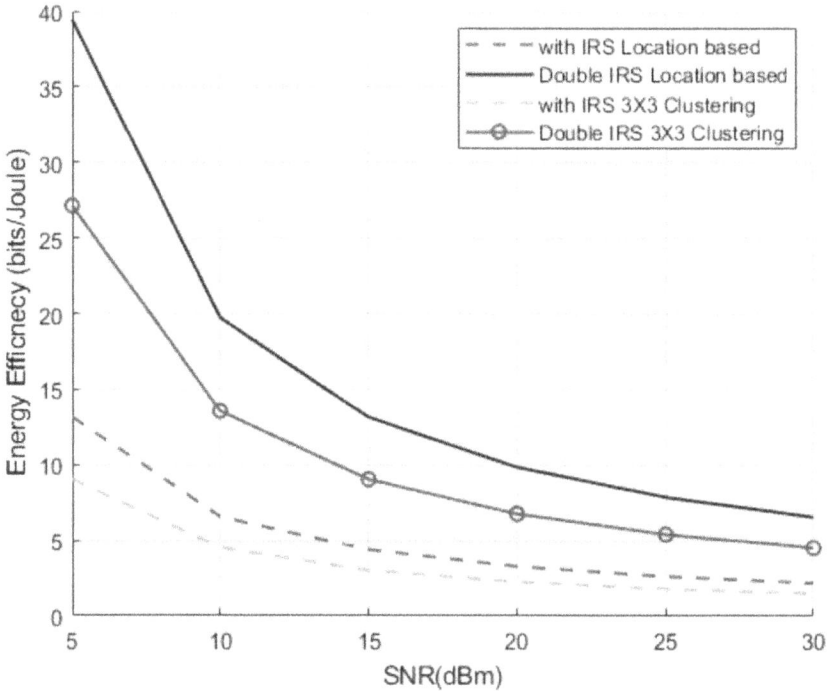

Figure 5.6 EE vs SNR with double IRS blocks.

5.5 IMPLEMENTATION OF IRS: CHALLENGES AND RESEARCH DIRECTIONS

It is also unclear how to operate the IRS in the THz transmission, despite the fact that the IRS's power usage is frequently undermined in the present IRS-assisted communications. The IRS's energy consumption and reflective coefficients must be considered in their entirety. The assumption used by traditional methods to get IRS channel estimation is likely that only one reflection component is ever active and the others are all turned off. It is necessary to reduce the key issues that were just mentioned. And it is difficult to use an element-by-element ON-OFF-based channel scattering architecture for a broad IRS with a large number of reflection components. Most of the time, only simulation studies are used to justify previous works on IRS. Further investigation is thus required to evaluate the theoretical findings with the data gathered by the practical and legitimate deployment. In addition, novel initiatives should be taken into the IRS discrete phase shifts problem in order to strengthen the applications of IRS-aided wireless networks. However, there are a few unresolved issues, hurdles, and other difficulties that have been described below.

5.6 CURRENT CHALLENGES AND RESEARCH DIRECTIONS

- *IRS deployment:* The distance between the IRS and the user in case of wireless communications may change; hence, we need to determine the exact location to place the IRS block. In addition, before deploying the IRS one needs to know the number of IRS components is enough to satisfy the users as well as the size of the IRS block with the desired location.
- *DL Training for IRS:* DL-based training for IRS requires a lot of data and a lengthy learning method. Finding the best channel conditions with strong connectivity and quick links and disabling difficulties with user movement is necessary because of the vast IRS components and huge users. The majority of the activities, such as how fast to educate the networks, how thoroughly to build the infrastructure, and what kind of environment to develop for the reliable and secure execution of communication, must be performed by the DL models. This causes a large transmission delay as well as security issues.
- *Beamforming:* Communication networks beyond 5G will also function at higher frequencies in the millimeter-wave spectrum. The installation of an IRS will enable the receiver to receive the required signal intensity in wireless networks. It is necessary to improve IRS phase-shift and beamforming at the transmitter. To determine the best combined active and passive beamforming approaches, more study is needed.
- *IRS phase shift:* Because of passive behaviors, many reflection components in real-world settings are unable to reflect the incoming signal in the required direction. This makes it more challenging to implement current signal processing techniques. Therefore, further research is required to identify significant control over the phase shift of the IRS components.
- *Channel Estimation:* Passive components play a large role in cellular connections with IRS. Therefore, using signal processing techniques to determine the channel estimate is a challenging task. In order to obtain the channel state information and manage the network connectivity without using power amplifiers, an innovative approach must be put into practice. Recently, channel estimation utilizing DL/ML-based algorithms for communication systems has shown superior results.

5.7 CONCLUSION

This chapter examines the EE of the ITS with IRS assistance under different scenarios. Since its introduction, ITS has experienced fast growth, as a result of its contribution to smooth data transfer among vehicles, safe transportation, and accident avoidance. However, the volume of data handled by ITS necessitates greater transmission power. IRS blocks with a number of passive reflecting components have been identified as a potential technique to overcome this issue and improve EE. To address the requirements for the actual implementations of IRS-assisted ITS, various fading environments have been adopted. Additionally, each IRS element's phase-shift optimization becomes complex, making it too difficult to estimate the channel for IRS-assisted ITS. Based on the desired location of the ITS, and unique IRS element clustering approach and a passive beamforming strategy to address the aforementioned issues and improve the EE and spectral efficiency of the IRS-assisted ITS with numerous IRS blocks are also examined. The performance of the ITS in terms of EE showed more improved performance, according to numerical findings, by the use of double IRS blocks.

REFERENCES

Abdullah, Z., Chen, G., Lambotharan, S., & Chambers, J. A. (2021). Optimization of intelligent reflecting surface assisted full-duplex relay networks. *IEEE Wireless Communications Letters, 10*(2), 363–367. 10.1109/lwc.2020.3031343

Alrabeiah, M., Demirhan, U., Hredzak, A., & Alkhateeb, A. (2020). Vision aided urll communications: Proactive service identification and coexistence. *2020 54th Asilomar Conference on Signals, Systems, and Computers.* 10.1109/ieeeconf51394.2020.9443526

Bjornson, E., & Sanguinetti, L. (2019). Demystifying the power scaling law of intelligent reflecting surfaces and metasurfaces. *2019 IEEE 8th International Workshop on Computational Advances in Multi-Sensor Adaptive Processing (CAMSAP).* 10.1109/camsap45676.2019.9022637

Bjornson, E., Ozdogan, O., & Larsson, E. G. (2020). Intelligent reflecting surface versus decode-and-forward: How large surfaces are needed to beat relaying? *IEEE Wireless Communications Letters, 9*(2), 244–248. 10.1109/lwc.2019.2950624

Chen, J., Liang, Y. C., Pei, Y., & Guo, H. (2019). Intelligent reflecting surface: A programmable wireless environment for physical layer security. *IEEE Access, 7*, 82599–82612. 10.1109/access.2019.2924034

Chen, Y., Wang, Y., Zhang, J., & Renzo, M. D. (2021). QoS-driven spectrum sharing for reconfigurable intelligent surfaces (RISS) aided vehicular networks. *IEEE Transactions on Wireless Communications, 20*(9), 5969–5985. 10.1109/twc.2021.3071332

Cheng, Y., Li, K. H., Liu, Y., Teh, K. C., & Vincent Poor, H. (2021). Downlink and uplink intelligent reflecting surface aided networks: NOMA and OMA. *IEEE Transactions on Wireless Communications, 20*(6), 3988–4000. 10.1109/twc.2021.3054841

Dampahalage, D., Shashika Manosha, K. B., Rajatheva, N., & Latva-aho, M. (2020). Intelligent reflecting surface aided vehicular communications. *2020 IEEE Globecom Workshops (GC Wkshps).* 10.1109/gcwkshps50303.2020.9367569

Dey, K. C., Rayamajhi, A., Chowdhury, M., Bhavsar, P., & Martin, J. (2016). Vehicle-to-vehicle (V2V) and vehicle-to-infrastructure (V2I) communication in a heterogeneous wireless network – performance evaluation. *Transportation Research Part C: Emerging Technologies, 68,* 168–184. 10.1016/j.trc.2016.03.008

Di Renzo, M., Ntontin, K., Song, J., Danufane, F. H., Qian, X., Lazarakis, F., De Rosny, J., Phan-Huy, D. T., Simeone, O., Zhang, R., Debbah, M., Lerosey, G., Fink, M., Tretyakov, S., & Shamai, S. (2020). Reconfigurable intelligent surfaces vs. relaying: Differences, similarities, and performance comparison. *IEEE Open Journal of the Communications Society, 1,* 798–807. 10.1109/ojcoms.2020.3002955

Dimitrakopoulos, G., & Demestichas, P. (2010). Intelligent transportation systems. *IEEE Vehicular Technology Magazine, 5*(1), 77–84. 10.1109/mvt.2009.935537

Ding, Z., & Vincent Poor, H. (2020). A simple design of IRS-NOMA transmission. *IEEE Communications Letters, 24*(5), 1119–1123. 10.1109/lcomm.2020.2974196

Duan, W., Gu, J., Wen, M., Zhang, G., Ji, Y., & Mumtaz, S. (2020). Emerging technologies for 5G-IOV networks: Applications, trends and opportunities. *IEEE Network, 34*(5), 283–289. 10.1109/mnet.001.1900659

Duan, W., Gu, X., Wen, M., Ji, Y., Ge, J., & Zhang, G. (2022). Resource management for itelligent vehicular edge computing networks. *IEEE Transactions on Intelligent Transportation Systems, 23*(7), 9797–9808. 10.1109/tits.2021.3114957

Hu, S., Rusek, F., & Edfors, O. (2018). Beyond massive MIMO: The potential of data transmission with large intelligent surfaces. *IEEE Transactions on Signal Processing, 66*(10), 2746–2758. 10.1109/tsp.2018.2816577

Huang, C., Zappone, A., Alexandropoulos, G. C., Debbah, M., & Yuen, C. (2019). Reconfigurable intelligent surfaces for energy efficiency in wireless communication. *IEEE Transactions on Wireless Communications, 18*(8), 4157–4170. 10.1109/twc.2019.2922609

Kong, L., He, J., Ai, Y., Chatzinotas, S., & Ottersten, B. (2021). Channel modeling and analysis of reconfigurable intelligent surfaces assisted vehicular networks. 10.1109/iccworkshops50388.2021.9473681 2021 IEEE International Conference on Communications Workshops (ICC Workshops)

Lu, Y. (2021). Intelligent reflecting surface aided wireless powered mobile edge computing. 10.32657/10356/164981

Lyu, B., Hoang, D. T., Gong, S., Niyato, D., & Kim, D. I. (2020). IRS-based wireless jamming attacks: When jammers can attack without power. *IEEE Wireless Communications Letters, 9*(10), 1663–1667. 10.1109/lwc.2020.3000892

Ozdogan, O., Bjornson, E., & Larsson, E. G. (2020). Intelligent reflecting surfaces: Physics, propagation, and pathloss modeling. *IEEE Wireless Communications Letters, 9*(5), 581–585. 10.1109/lwc.2019.2960779

Pan, C., Ren, H., Wang, K., Xu, W., Elkashlan, M., Nallanathan, A., & Hanzo, L. (2020). Multicell MIMO communications relying on intelligent reflecting surfaces. *IEEE Transactions on Wireless Communications, 19*(8), 5218–5233. 10.1109/twc.2020.2990766

Pegorara Souto, V. D., Souza, R. D., Uchoa-Filho, B. F., Li, A., & Li, Y. (2020). Beamforming optimization for intelligent reflecting surfaces without CSI.

IEEE Wireless Communications Letters, 9(9), 1476–1480. 10.1109/lwc. 2020.2994218

Peng, Z., Zhang, Z., Pan, C., Li, L., & Swindlehurst, A. L. (2021). Multiuser full-duplex two-way communications via intelligent reflecting surface. *IEEE Transactions on Signal Processing, 69*, 837–851. 10.1109/tsp.2021.3049652

Perovic, N. S., Tran, L. N., Di Renzo, M., & Flanagan, M. F. (2021). Achievable rate optimization for MIMO systems with reconfigurable intelligent surfaces. *IEEE Transactions on Wireless Communications, 20*(6), 3865–3882. 10.1109/twc.2021.3054121

Rajak, S., Muniraj, I., Elumalai, K., Hosen, A. S., Ra, I. H., & Chinnadurai, S. (2022). Energy efficient hybrid relay-IRS-aided wireless IOT network for 6G communications. *Electronics, 11*(12), 1900. 10.3390/electronics11121900

Shaikh, M. H., Bohara, V. A., Srivastava, A., & Ghatak, G. (2021). Performance analysis of intelligent reflecting surface-assisted wireless system with non-ideal transceiver. *IEEE Open Journal of the Communications Society, 2*, 671–686. 10.1109/ojcoms.2021.3068866

Smith, B. L., & Venkatanarayana, R. (2005). Realizing the promise of intelligent transportation systems (ITS) data archives. *Journal of Intelligent Transportation Systems, 9*(4), 175–185. 10.1080/15472450500237288

Song, W., Rajak, S., Dang, S., Liu, R., Li, J., & Chinnadurai, S. (2022). Deep learning enabled IRS for 6G intelligent transportation systems: A comprehensive study. *IEEE Transactions on Intelligent Transportation Systems*, 1–18. 10.1109/tits.2022.3184314

Taha, A., Alrabeiah, M., & Alkhateeb, A. (2021). Enabling large intelligent surfaces with compressive sensing and deep learning. *IEEE Access, 9*, 44304–44321. 10.1109/access.2021.3064073

Wang, Y., Lu, H., & Sun, H. (2021). Channel estimation in IRS-enhanced mmwave system with super-resolution network. *IEEE Communications Letters, 25*(8), 2599–2603. 10.1109/lcomm.2021.3079322

Wang, Z., Liu, L., & Cui, S. (2020). Channel estimation for intelligent reflecting surface assisted multiuser communications: Framework, algorithms, and analysis. *IEEE Transactions on Wireless Communications, 19*(10), 6607–6620. 10.1109/twc.2020.3004330

Wu, Q., & Zhang, R. (2019). Intelligent reflecting surface enhanced wireless network via joint active and passive beamforming. *IEEE Transactions on Wireless Communications, 18*(11), 5394–5409. 10.1109/twc.2019.2936025

Wu, Q., & Zhang, R. (2020). Beamforming optimization for wireless network aided by intelligent reflecting surface with discrete phase shifts. *IEEE Transactions on Communications, 68*(3), 1838–1851. 10.1109/tcomm.2019.2958916

Wu, Q., Guan, X., & Zhang, R. (2022). Intelligent reflecting surface-aided wireless energy and information transmission: An overview. *Proceedings of the IEEE, 110*(1), 150–170. 10.1109/jproc.2021.3121790

Yang, Z., Chen, M., Saad, W., Xu, W., Shikh-Bahaei, M., Poor, H. V., & Cui, S. (2022). Energy-efficient wireless communications with distributed reconfigurable intelligent surfaces. *IEEE Transactions on Wireless Communications, 21*(1), 665–679. 10.1109/twc.2021.3098632

Yu, G., Chen, X., Zhong, C., Ng, D. W., & Zhang, Z. (2020). Design, analysis, and optimization of a large intelligent reflecting surface-aided B5g cellular internet of things. *IEEE Internet of Things Journal*, 7(9), 8902–8916. 10.1109/jiot. 2020.2996984

Zhang, J., Wang, F. Y., Wang, K., Lin, W. H., Xu, X., & Chen, C. (2011). Data-driven intelligent transportation systems: A survey. *IEEE Transactions on Intelligent Transportation Systems*, 12(4), 1624–1639. 10.1109/tits.2011. 2158001

Zheng, B., Wu, Q., & Zhang, R. (2020). Intelligent reflecting surface-assisted multiple access with user pairing: NOMA or OMA? *IEEE Communications Letters*, 24(4), 753–757. 10.1109/lcomm.2020.2969870

Zhu, Y., Mao, B., Kawamoto, Y., & Kato, N. (2021). Intelligent reflecting surface-aided vehicular networks toward 6G: Vision, proposal, and future directions. *IEEE Vehicular Technology Magazine*, 16(4), 48–56. 10.1109/mvt.2021. 3113890

Chapter 6

AI applications at the scheduling and resource allocation schemes in web medium

R. Shekhar[1], P. Mano Paul[1], and Diana Jeba Jingle[2]

[1]Department of Computer Science and Engineering, Alliance University, Bangalore, India

[2]Department of Computer Science and Engineering, Christ University, Bangalore, India

6.1 INTRODUCTION

Human communication has improved and increased significantly due to technology. Almost 2 billion individuals have access to information, multimedia, messaging, audio and video chats, and many other distinct features, which can be attributed to the technological innovation in wireless communication. Because of this situation (Younge et al., 2010), identified the network, expansion over the number of internet-connected, technology-dependent smart devices has expanded and is continuing to grow.

The demand for massive processing capacity, information storage, and high-speed broadband has skyrocketed with growth in the use of IoT applications that enable real-time decision-making systems to stream data, as suggested in Jingle & Paul (2021). Cloud computing, which is now being complimented by fog computing, has already tried to meet these needs. Large data streams are handled by cloud and fog computing, facilitating, as well as a vast number of IT-based applications. IoT devices benefit greatly from cloud computing, yet cloud computing has several difficulties when working with IoT devices which were described in Kumar et al. (2020). These difficulties include problems with latency, mobility support, insufficient network bandwidth, and location awareness. Here the technology called "fog computing" is now available to enable cloud computing to address these issues with an open application programming interface suggested by Baccarelli et al. (2017), fog computing approach makes conducting research more convenient than batch processing. The cloud layer is made up of numerous storage systems and powerful servers that offer various application services. It holds many data and supports intensive computing and analysis. Network fog nodes such as base stations, switches, routers, and gateways make up the fog layer. These nodes assist in computation, transmission, and temporary data storage. This layer supports applications that require low latency.

DOI: 10.1201/9781003369028-6

Fog computing is the term used to describe the extension of cloud computing (CC) to accommodate the needs of cutting-edge innovations such as 5G (Hassan et al., 2019), the Internet of Things, artificial intelligence, and other similar technologies. Data processing and storage services are among the offerings made by fog computing (FC) to IoT consumers. By storing the data locally on a fog node rather than transmitting it to the cloud for storage, fog computing expands its offerings to include cloud computing, increasing the effectiveness and performance of the cloud in this way. Fog computing slows down the rate at which data must be processed and transported to the cloud, making it easier for the cloud to store and analyze data. This assistance of processing leads to decreasing network traffic, delays, and so on.

In Khan et al. (2020) introduces a reinforcement learning (RL) method for resource provisioning methods. In the RL-based approach, components are classified into the knowledge base and decision-making process. A population-based optimizer is qualified to offer knowledge base actions. To make the ultimate judgments on resource usage and resource limitations, an action selection system is employed. A learner system is used to gather data regarding relationships between system states, actions, and responses. The recommender system uses a supervised learning procedure for offering an action based on the state specified with immediate reward and cumulative reward, used to find the best action in the possible action domain.

Cloud computing benefits from fog computing, given the rise in IoT devices, various issues must be resolved to improve fog computing's effectiveness and efficiency given according to Dabbagh et al. (2015). It includes the ground-breaking idea of giving IoT devices access to storage and computational power. It is made up of the following components in that order: IoT devices (lower tier), fog-nodes (middle tier), and cloud (top tier). The IoT layer (Ajay et al., 2021) links any physical thing, including household appliances, wearable technology, cameras, and car sensors, producing enormous amounts of data. Support for information inquiry and capacity frameworks is provided at the foundational level via the cloud. Most IoT applications require less latency. Cloud service providers are focusing on how to increase return on investment (ROI) such as by lowering the total cost of ownership (TCO), improving service quality, accelerating time to value, and supervising the intricate values in the cloud.

Fog computing suggested by Dastjerdi et al. (2016), allows services to be offered outside the cloud and reduces network congestion and latency as a result. The three-tier framework is depicted in Figure 6.1. Build hosts (input, virtual machines, cost per storage & OS), storage list, set parameters of upper bandwidth, lower bandwidth, latency, and request mapping to modules have all been used to develop the fog node design. Challenges of fog computing include security, scalability, resource management, energy, consumption, latency, heterogeneity, and dynamicity. The device layer is made up of many intelligent devices that make it easier to recognize feature data in actual objects. The device layer also distributes data for processing and storage.

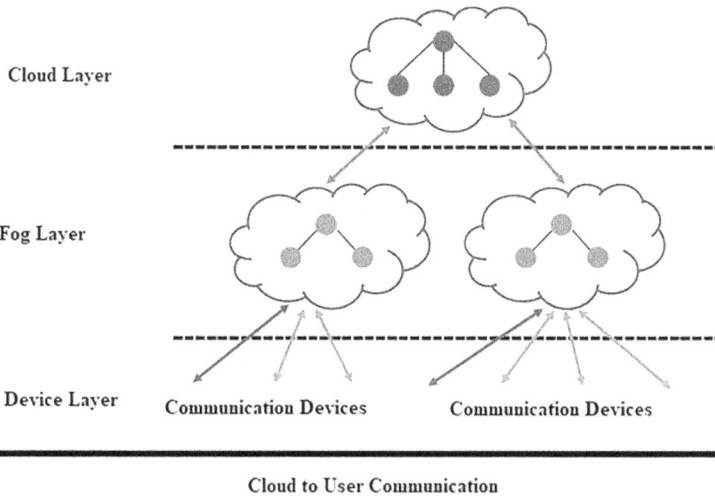

Cloud Layer

Fog Layer

Device Layer   Communication Devices   Communication Devices

Cloud to User Communication

Figure 6.1 Cloud-device layered architecture.

The term "autonomic computing" refers to the central nervous system of our bodies, which adjusts to various states of affairs spontaneously and without outside assistance in various states (Qu et al., 2009). The human body may sweat if it is under pressure; shivering is a sign that the human body is cold. Without outside assistance, the human body does these activities automatically. Like this, autonomic computing is used to tackle complicated issues in cloud architecture. Autonomic computing is used to develop intelligent systems that can monitor all complicated problems, identify themselves, constantly tune themselves, adjust to erratic situations, prevent, and recover from failures, and provide a secure environment. In the realm of IT, autonomous computing is not a sudden revolution, and it will be performed with user devices, as given in Figure 6.2. Autonomic computing will result in several changes in both hardware and software. Different transition levels focused are listed below:

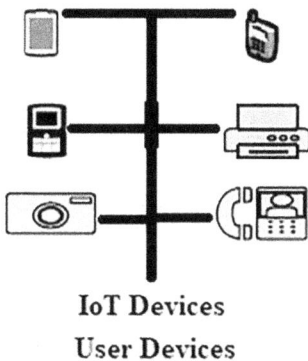

IoT Devices
User Devices

Figure 6.2 User devices connected with fog nodes.

6.1.1 The first level: Basic

It is an example of labor-intensive computing, where each component is installed and worked on separately. It needs highly qualified individuals to collect and analyze the generated data.

6.1.2 Second: Managed

It can be accomplished by gathering data and taking action with the use of management tools, generated data is thoroughly examined, and then acts. This system has the benefit of having excellent system awareness and productivity.

6.1.3 Predictive at the third level

The system investigates and correlates the data at this level. The cloud provider authorizes and starts the actions.

Organizational integration across many components is piled up at this level. This decreased the dependence on deep skills.

6.1.4 The fourth level: Flexible

At this stage, the system organizes into groups and creates action plans based on the policies. It enables the workforce to achieve performance success relative to the goals. A balance between system and person-to-person interactions is beneficial for an organization. It aids businesses in adapting to shifting commercial demands.

6.1.5 Fifth level: Autonomic

The elements at this level are closely interconnected and adapt to the demands of the business. It enables people to concentrate on pressing business needs. In the current IT industry, it becomes new business model. Autonomic computing must accomplish business agility and resilience.

In the rest of the article is based on the architectural blocks in wide area metropolitan area networks. We refer to them commonly as MAN. This includes WiMax, LTE, CDMA, and others.

6.2 RELATED WORK

Load balancing in a fog environment is a key element that considers resource allocation, performance, and other management metrics involved in load balancing. According to the study, the author (Singh et al., 2015) identifies problems with load balancing that arise with various applications, and the software is divided into smaller jobs, which are then finished concurrently. They are to conduct the required operation using distributed and parallel computer systems. The researchers suggest using evolutionary algorithms to construct load arrangements as a solution to the dynamic load balancing issue. Particle swarm optimization (PSO) was considered in VMs for attaining load balancing in a study by (Sharma et al., 2018) producing optimized results. An analysis of the demand placed on the servers, and new suitable methods were planned, including swarm intelligence that is used for load-balancing optimization, continuing in the same vein as (Chitra Devi et al., 2015) for other computing. Another optimization problem is NP, which is a challenging issue, and considers the underloaded condition of the VM through this approach. Also, another suggestion was made by maximizing the QoS service parameters and throughput, in the load-balancing strategy for virtualization approaches due to its complexity.

The author named Baccarelli et al. (2017) described a paradigm for fog computing (FC) that makes use of cloud atomization technology. The author states that the load balancing is done using a dynamic graph partitioning theory to develop load-balancing methods in fog computing. With less node ingestion over fog nodes, the load-balancing process has effectively ordered the system's resources. According to Chitra Devi et al. (2015), a well-organized technique for load balancing was introduced in a fog computing environment. This satisfies user expectations by maintaining data consistency with fewer hassles, increasing throughput, using the network, and time-efficiently scheduling jobs. Problems related to load balancing and scheduling of tasks, in a post-planned scenario as per requirement resource allocation, should be done directly this is known as "auto-allocation". For fog schedulers to be assured about the successful usage of resources to be efficient, and the incoming workload should be made resource allocated (Baccarelli et al., 2017) through fog nodes, the task allocation to nearby processors and scheduling techniques needs to be improved. For greater resource use in FC, effective resource management is crucial. Here numerous distinct kinds of obstacles in FC.

- *Fluctuating and Dynamic Workload:* Fog computing experiences its changes in two different ways: preset and post-determined. Because of an earlier analysis of a preset fluctuation condition, resources are allocated effectively and within the allotted time frame.
- *Ensuring Effective Resource Use:* "Auto-allocation" refers to the requirement-based direct resource allocation that should occur in a post-planned scenario. The incoming workload should be resource allocated to provide fog schedulers confidence in the efficient use of resources. Scheduling methods need to be improved for work distribution to neighboring processors.

- *Heterogeneous Physical Nodes in Fog Data Centers:* Tasks are scheduled inside the reachable nodes. These nodes are dispersed across several architectures and geographic regions, with varying memory, network performance, and computational capacity.
- *More Scheduling Granularity than Current Scheduling:* From cozy, informal scheduling, the challenge of scheduling has risen in magnitude.

Load balancing in fog environment nodes is used in its applications to manage the resource requirements. By managing load balancing, these resources can be acquired and used more effectively. Use effective load balancing to avoid factors such as low load, bottlenecks, and overload that interfere with the effective use of resources. Load balance continues to be difficult to implement in fog situations when running its applications. Therefore, to address this issue, the current part of the research is devoted to applying an optimization technique to identify an effective load balancing solution. Load balancing and energy consumption proceed with the two most pressing problems for fog computing. The technique of effectively spreading inbound network traffic among the backend servers or the server pool is known as load balancing. The main needs of consumers are computation and storage. This workload at fog nodes should be considered when allocating resources in a fog environment. Moreover, resource managers and schedulers must be used to determine the ideal fog nodes (Khosravi & Buyya, 2018), resulting in a load imbalance issue. Some of them are left inactive, while others become overburdened in this situation. In fog computing, there are three steps in the load-balancing process. A load balancer is used in the initial stage to gather data on the workload distribution. This decision-making process for the most likely data distribution is the focus of the phase. The third stage is where data is finally transferred from one overloaded node to another. Fog computing provides a solution for cloud infrastructure because of this ever-rising energy usage. For the community of contemporary researchers, the reduction in energy usage through fog computing has been a difficulty.

Increased energy consumption is a brand-new problem that has emerged because of the increased use of cloud computing. The number of PCs is rising in cloud data centres (CDC) to improve processing and efficiency; energy use increases because of this. Resource allocation techniques via requests of the users is an aspect on which the consumption of energy is dependent across servers (Lin, 2012). Thus, it is important to develop an energy-aware scheduling process to save on energy in the fog computing environment as well. Based on this discussion (Sharma & Awasthi, 2018), the present research is focused on presenting the solutions to issues of load balancing and energy consumption along with cloud nodes.

Here, different researchers use different scopes with different methodologies to deal with the cloud layer and fog layer. A cloud layer with no latency issue uses PSO, ant colony optimization, and multi-agent generic algorithm methodology. The focus on these issues is to rebuild load balancing in cloud layers and to provide reliability, by building their application with task scheduling and achieved in a genetic algorithm and limited with few resources as said (Nikravesh et al., 2017) in energy and policies. Here is another application discussing cost optimization with particle swarm algorithm and few of the energy resources and data centers will be limited for sharing resources. These metrics follow the cloud layer and with the fog layer, few techniques lagged to perform multitasking toward load balancing and scheduling, and its time needs to be reduced.

6.3 OBJECTIVES OF THE STUDY AND STRATEGIES

A "Secure Delay Aware Scheduling and Load Balancing for Deadline-Sensitive Applications in Fog Computing Environment" is what the project attempts to create. The following is a definition of these objectives:

- To create a framework for scheduling applications with tight deadlines.
- Making use of optimization techniques to deal with the load-balancing problem.
- Minimizing service-level violations and managing energy usage in fog computing.
- Using a newly created four-tier architecture in a fog environment to address the problems of load balancing and delay aware scheduling.
- To offer secure deduplication and fog-aided work allocation for cluster-based industrial IoT.

The resource is an economic or productive factor needed to achieve an activity. Resource allocation is a scheduling of activities and is a part of resource management which was dealt with using a multi-layered approach. In the resource allocation process, the available resources are assigned with basic allocation decisions, contingency mechanisms, and planning and assessment approaches are integrated with each other for allocating the resource. Space assignment, an establishment quality-of-service (QoS) criteria such as delay, delay jitter, and throughput requirements are satisfied.

Resource allocation for MAN has numerous strategies (Hussain et al., 2019) via quality of service and coverage of cellular networks (Paul & Ravi, 2018), which were discussed and integrated into large-scale networks for online communication channels. In MAN systems, the resource is allocated to various mobile stations in both downlink and uplink channels. The base station (BS) based on the burst size chooses an optimal data packet from the packets waiting in the queue. The decided packet is transmitted to various subscribers. Depending on the data rate of the packet, a certain quality of resources is allocated to the cluster. The resource allocations are mapped into a MAN frame in a rectangular region to achieve the QoS requirements. The resource scheduler allocates the resources for the burst mapping algorithm. The energy consumption of a mobile station (MS) is minimized for resource mapping. The MS is connected to different wireless access via MAN networks. IP-based mobile broadband allows compatibility with multiple internet applications. The mobile MAN increases the capacity and coverage by using multiple-input multiple-output technology and adaptive antenna systems. The convergence of the mobile and fixed broadband networks is enabled through a common wide-area and flexible network architecture. The features of the mobile MAN are high data rates, quality-of-service (QoS), scalability, security, and mobility.

By leveraging adaptive antenna systems and multiple-input multiple-output technology, MAN expands both capacity and coverage. A single wide-area and flexible network design enables the fusion of mobile and fixed broadband networks. High data speeds, quality-of-service (QoS), scalability, security, and mobility are some of the characteristics of MAN. The major features of end-to-end MAN architecture are scalability, extendibility, coverage, operator selection, multi-vendor interoperability, and the QoS. The QoS simultaneously enables flexible support of IP services and supports admission control and bandwidth management.

6.3.1 Algorithm for load balancing methods using priority ordering

Function priority order (Load, count):

1. Initialize the load be 0
2. For every job, identify a connect as var count
3. For each valid count, calculate the load
   a. Load = existing load + new weighted factor in load
4. Update the steps for each stage of the job assigned with its weight factor
5. Until all factors for weight are assigned, add the cost such that w<=0
6. Energy for each node can be calculated as $\sum$ Every node weightage to its total count
7. New weighted count * random() < Total count (Load) * random()

Through service-level agreement (SLA) in fog environment, we can effectively utilize resources which have been provisioned, shared, monitored, allocated, and executed as well to predict the quality of service within the knowledge pool. It consists of a local module and a remote module to access the resources with different computational environments for their resource allocation and usage factors. Here different models were also performed using fog computing.

- *Heterogeneous Resource Allocation techniques:*

a. Simple allocation methods
b. Method of a grouping of activities

- *Simple Allocation methods:* Various activities are assigned to share resources with different types. Here, a plan of activities contains {1 to 11} in which resources will be allocated to Q.

Let it be assigned as Q = {T1, T2, T3}
T1 = {5,10,6,7,8,9}
T2 = {11,2}
T3 = {1,3,4}

Here, the next step is to group the activities, which is a subset of activities that can be eliminated from problem P in such a way that it provides an optimal solution to the original problem that can be obtained from an optimal solution to the reduced problem of p1. Here, a grouping of activities means clustering into a single activity which is termed a macro activity which is a possible set of original activities. A sequence of activities is formed by activities that do not overlap with one another and are processed with the exact resources. If two activities produced the same sequence, then clustering performs until the end of the macro activity, which coincides with the resource units.

6.3.2 Coverage-based cell selection algorithm

The 6G networks are provided a QoS and high-security end-to-end delay. The seamless connection of the 6G network is provided to huge services and retrieving information, images, and videos with rechargeable energies to execute the devices. The key features of the 6G are highly dynamic and application adaptability. Advantages of the 6G wireless systems:

- Support audio, streaming video, internet, and other broadcast services in addition to interactive multimedia;
- Offers minimal cost, high capacity, and great speed;
- Better scheduling and call-admission control methods are provided;
- Offers improved spectral efficiency; and
- Provides scalable mobile services, worldwide access, and service portability.

```
┌─────────────────────────────────────────┐
│                Services                   │
└─────────────────────────────────────────┘
        │
        ▼
┌──────────────────────────┐    ┌─────────────────┐
│   Control Management      │───►│                 │
└──────────────────────────┘    │     Network     │
        │                       │                 │
        ▼                       │                 │
┌──────────────────────────┐    │                 │
│      Core Network         │───►│                 │
└──────────────────────────┘    └─────────────────┘
        │                               │
        ▼                               ▼
┌─────────────────────────────────────────┐
│   Connectivity Servers and Databases     │
│   With its rechargeable Power Supply      │
└─────────────────────────────────────────┘
```

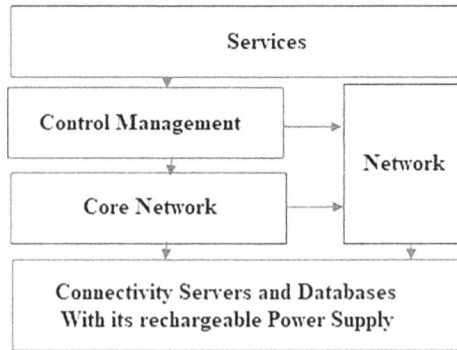

Figure 6.3 Next-generation network structure.

The 6G wireless system is a high-level Wi-Fi with a rechargeable wireless system with high throughput and extensive coverage. The high spectral efficiency is intended to be provided by it. Also, the network structure for the next generation 6G communication system is given in Figure 6.3.

MAN provides a wireless communication technology to provide high data-rate communications in metropolitan area networks through global interoperability. The MAN core network's architecture includes optical integrative switching and an all-IP mobile network, routers, a location registry, a home agent, a proxy server, and an interworking gateway in the MAN connectivity service network. Domain name server uses a dynamic host configuration protocol (DHCP) to identify IP dynamically towards domain server, and the user database that is externally connected over the internet. A universal serial bus (USB) provides connectivity to the MAN network for digital systems. The combination of network functions is used to define the functionalities and allocate the mobile station and endpoint parameters for user sessions using MAN networks. Inter-autonomous systems provide mobility and connectivity to MAN services i.e., location-based services and USB provides connectivity to the MAN network, which enlarges the QoS class for VoIP applications.

6.3.3 Cell degree-based resource allocation (CBRA)

The call admission control (CAC) plays a vital role in QoS provisioning that mentions a network QoS mechanism. A new connection with the QoS requirements is established by determining the network QoS. After the registration process, a sub-system is connected to the base station because the MAN is connection oriented. A CAC module is used to decide whether the connection is accepted or rejected. The base station grants bandwidth in two ways, such as grant per connection (GPC) and grant per subscriber station (GPSS). It is scalable because less information is transmitted to the base station. A critical section for the effective design, deployment, and development of broadband wireless networks is cell planning. A severe process of investigation, the definition of system-wide technology parameters, and analysis and allocation of radio frequency resources are used for efficient planning. A cell selection algorithm is introduced in this research. Cell planning is performed in every network to improve the coverage of the BS and to avoid interference. The major objective of cell planning is to improve the traffic capacity and performance by minimizing the system coverage and spectrum efficiency. The performance of this proposed coverage-based cell selection (CBCS) algorithm is compared with the performance of the existing approaches such as the greedy algorithm, bounded greedy weighted algorithm, call admission control, and joint scheduling and resource allocation. Resource allocation and scheduling are used to allocate sub-carriers and power to the user. The service requirement of the user is achieved while maintaining user fairness and improving resource usage. An opportunistic scheduling scheme is used to compensate for the user's unfairness of the scheduling. A scheduling scheme is used in orthogonal frequency division multiple access (OFDMA) networks to protect user fairness.

A practical scheme is efficiently developed for the medium access control (MAC) layer of the OFDMA networks with low complexity. The scheduling process is performed to achieve the optimal usage of the resources to satisfy the QoS guarantees, improve throughput, and reduce power consumption while securing feasible algorithm complexity and system scalability. The guidelines for resource allocation are geographical criteria, thematic criteria, diversity of stockholders and shared governance, relevancy, and BNDES allocation lines. The performance of the MAN system depends on the improvement of network throughput and user fairness. The game theory algorithm is used to enhance the network throughput by guaranteeing user fairness. The channel status values of the relay links are calculated by receiving the signal from the base station and the mobile station. The calculated channel status values are compared with a preset reference value for selecting a relay mode of the relay station in a multi-hop-based network in a multi-hop network. The received signal is relayed for selecting the relay mode according to the results of the compared result. The complete performance of a MAN system depends on the augmentation of network throughput and user fairness. The cell degree is introduced to balance the connection between the system network throughput and user fairness.

The cell degree in the multi-hop network is depending on the user accessing the data rate and the QoS parameter. The throughput of the proposed CBRA is high compared to the existing ARM. In this proposed CBRA, the channel quality of the cumulative distributive function (CDF) is compared with the quality of the signal-to-interference-plus-noise ratio (SINR). The UE and the BS are used to evaluate the data rate and QoS parameters such as bandwidth, latency, jitter, cell loss ratio, and cell error ratio. The system network throughput, the user fairness, and the game theory are incorporated with each other to provide a higher throughput and bandwidth consumption ratio than the existing greedy approach by using the game theory and the CBRA algorithm.

A utility maximization framework for a discrete allocation originates to balance system efficiency and user fairness. In adaptive scheduling, packets are transmitted depending on allotted slots from various priorities of traffic classes based on the channel condition. Routing and scheduling bandwidth allocation problems are solved by centralized routing and centralized scheduling (CRCS). The traffic and position information of the stations are received by BS, which transmits the corrected routes and returns them to the substation. In distributed routing and centralized scheduling, the routes of the subsystem are established at the base station. The base station receives information about the subsystem after finding the route. The base station determines a collision-free transmission schedule.

6.3.4 Distributed routing and scheduling techniques

Routing algorithms for centralized scheduling are used to accommodate network interference and load balancing. The scheduling procedure makes use of link scheduling and channel assignment algorithms. The QoS is ensured via a thorough architecture for MAN networks, which provisioned MAN utilized for the support, operability, accessibility, and retain the ability, integrity, and security that make up the QoS's core elements. Throughput, average latency, average jitter, and packet loss are presented as MAN metrics. Source processing delay, propagation delay, and destination processing delay are the main causes of latency. The network's consistency and stability are assessed using the jitter. The following ideas form the foundation of the handover algorithm whenever a new user is added.

A round-robin (RR) scheduling procedure and an analyzing process are run after switching to the new user. The signal to interference plus noise ratio's (SINR) limitations, after choosing the routes, are the connections made. On the routes, the flow rates are chosen. A scheduling procedure determines the time slot. By including the constraints to the input power level, the bi-criteria approximation problem is avoided, and the poly-log factor is guaranteed. According to the real load condition, the BS is overloaded. By using a link's SINR, a modulation and coding scheme (MCS) level can be calculated. The LB-based handover technique is used to guarantee the QoS and spread the traffic load. The BS employs a handover mechanism based on load balancing (LB).

The ARM is efficiently used for LB in the MAN network. Compared with the existing CBCS approach, the proposed ARM achieves better traffic management and load balancing without any channel estimation. To perform the channel estimation and to achieve user firmness, a cell-degree-based resource allocation (CBRA) scheme is proposed. In a distributed routing and distributed scheduling (DRDS), the station establishes its routes to the base station. In hybrid routing and scheduling (HRS), the routing and scheduling algorithms are implemented in a distributed way or centralized way. A CBRA mechanism in the MAN network uses both game theory and the CBRA approach. A user or BS can access the data rate as well as a QoS metric such as bandwidth, latency, jitter, cell loss ratio, and cell error ratio. To determine the cell degree, the user equipment (UE) and base station were compared. The BS and the user are provided with a MAN large-scale network architecture. The management of an IP address pool is offered through IETF standard techniques such as DHCP. A MAN network's primary components are its remote or mobile stations. The stations are situated inside the user's premises and come with the user equipment, whether fixed or mobile. MAN's network coverage area, known as the access service network (ASN), is a radio wave.

ASN is used with several base stations and many ASN gateways; the MAN network area creates a radio access network at the edge. Internet protocol (IP) connectivity and IP core network operations are made available to the MAN network through the connectivity service network (CSN). The CSN is a component of a home network service provider.

6.4 THE PROPOSED MODELS IN RESOURCE ALLOCATION

6.4.1 Load balancing as a resolution for fog computing

Load balancing is one of the most important problems with fog computing that needs to be addressed. Load balancing continues to be a significant difficulty even in a heterogeneous environment of fog computing when there are so many nodes and resources available to share the growing demand. This work has so concentrated on the minimization of the energy consumption rate and the lowering of the scheduled duration runtime. The current research has developed a method to obtain these findings. For load management in the fog computing environment, this approach uses artificial bee colony optimization (ABCO) to balance the load.

A crucial element that determines the effectiveness of resource allocation and management systems is load balancing. The problem of allocating workload that can be designed to cut down on power consumption and service latency for a given a job schedule in a cloud-fog scenario. In this study, we show that a fog provider can efficiently execute large-scale offloading applications by leveraging the link between their fog nodes and leased cloud nodes. As a result, a heuristic-based method is proposed, with its major objective being to establish a balance between the costs of cloud resources.

6.4.1.1 Reducing energy consumption and violation of SLA

The use of cloud computing has grown exponentially during the last decade. Due to the growing number of IoT devices, this trend is anticipated to expand much more in the upcoming years. Real-time tasks must be completed by change data capture (CDC), which is serverless data growth along with its usage. Due to insufficient resource bandwidth, typical data centers struggle to function in such a situation. Fog computing has been introduced to the globe as a remedy for this facet of cloud computing. It is also a supplement to cloud computing that has brought the cloud computing model out to the edge of the network. Hence, a wide range of innovative services and applications are based on infrastructure. A transitional technology between CDCs and IoT sensors or devices is fog computing. To improve the cloud-based service options for its sensors and devices, it offers storage services, networking, and computing. Energy consumption has increased because of the growth in so many facets of cloud computing. For researchers, cutting energy using fog computing has undoubtedly been a challenge. The factor on which the usage of energy across fog servers is dependent is resource allocation strategies for user requests. It allows for processing at the edge with a chance of cloud communication.

6.4.1.2 Delay-aware scheduling and load balancing: The solution in a four-tier architecture

The two key components of fog computing that have a substantial impact on an arrangement's performance are load balancing and scheduling. Numerous scholars have been working to develop a model that will enable job scheduling and load balancing to be done more effectively and efficiently. Many research efforts are currently being made to understand the scheduling idea in fog computing. A group of the fog nodes may bring these novel issues. In the fog environment, optimal task distribution or dynamic load balancing is a crucial problem. Nevertheless, the fog node is not balanced at the moment. Due to factors such as fog node failures, resource scarcity, and distributed or centralized environments, to name a few, the prior methodologies put out by other academics are trailing. The current study has been focused on reducing energy consumption and latency in the fog environment using both load balancing and effective job scheduling to address all the difficulties raised in the earlier studies. The current research has suggested a four-tier architecture to address the problems with task scheduling and load balancing observed in the earlier efforts.

The fog-computing environment consists of four separate tiers: 1 to 4.

By assessing the performance metrics for the VSOT application, the planned system has been validated:

- Reaction time for various task counts;
- Scheduling time/length for various job counts (seconds);
- Delay;
- Load-balancing rate; and
- Use of dnergy (KJ).

6.4.1.3 Task allocation and secure deduplication: Assistance from fog computing

Duplicate data is now being transmitted via the internet at a higher rate because of the proliferation of IoT devices. Duplicate data transfer has increased transmission, which has put more strain on data centre resources. The transmission of duplicate data results in the CS delaying the timely delivery of services to users. Here, a sizable quantity of unstructured data floating around in the world, including data that cannot be treated manually or by basic programs. Data are increased exponentially including the worldwide web, corporate services, applications, and networks, among others. Since this enormous expansion in data cannot be understood simply or automatically retrieved, planned awareness and information cannot be easily acquired. These factors contributed to the development of data mining and data science a well-known field that is pervasive in the current information age. The amount of data now is greater than what earlier processing systems could handle. The rise in new technologies, such as fog computing and the decline in hardware costs, are driving the exponential growth of online information. The data analytics community has a "big" problem with this scenario. Hence, "big data" is defined as a compilation of enormous volumes, velocities, and various data that must be processed quickly. Before the advent of big data, distributed computing has long been used with the use of numerous regular and persistent methods.

To examine the learning process, the researcher replaced their distributed versions. However, a distributed solution is now required for the majority of today's huge difficulties because architecture as a whole is unable to handle such significant problems. Over the past decade, numerous large-scale processing stages have attempted to address the big data problem. These phases make an effort to make distributed technologies more similar to standard users. These stages must be generated and maintained using complex methods, which makes the practice of mobile computing. Furthermore, platforms with several data require additional algorithms that finance associated operations such as big data pre-processing.

To investigate huge data sets, the author must redesign the typical algorithms for these activities. The primary framework that made it possible to handle big data sets was MapReduce. Large data sets were intended to be developed and produced using this ground-breaking tool. By using map and reduce, the employer can contemplate an accessible and widespread gadget without upsetting the technological nuances, such as failed retrieval, data separation, or job communication. The most common open-source implementation of MapReduce that maintains the fore-mentioned characteristics is Apache Hadoop. Hadoop and MapReduce are not meant to be measured because of their extreme adoration. Hadoop's replacement, Apache Spark, was touted as being capable of rapidly distributed computing via in-memory primitives. This was to store data in memory and reuse it regularly, which eliminates the problems with recitative and network processing that MapReduce presents. Moreover, Spark is a multipurpose framework that enables the execution of numerous distributed programming prototypes, such as Hadoop. A new abstraction module called "resilient distributed data sets" (RDDs) is made over the Spark. This versatile module enables, among other things, managing data segregation and perseverance control.

6.4.1.4 Data migration over cloud or fog based on applications

The "Cloud Adoption Toolkit" was provided in research [3] and [4] as a tool for comprehending the practicality of cloud migration. The upcoming factor has been taken into account:

1. The price of adopting cloud computing
2. Risk administration
3. Taking into account trade-offs between migration risks and cloud benefits.

The study's author has studied and observed the cutting-edge problem of viability and potential. The suggested cost modeling approach is cutting edge and is founded on current infrastructure expenses, minus usage. The capacity is to regulate the activity from power-generation technologies to prevent over-provisioning on the new cloud. The cost estimation tool displayed was known as the Plan for Cloud Globally. To estimate the operating costs related to the cloud, the authors advised displaying an application along with its fundamental substructure and organizational structure. A strategy for "enterprise applications" transfer to the cloud was suggested. The installation of a "cloud-permitted" version and representing the current system were their main priorities.

The distinguished strong suit involves

1. Advanced automation
2. The ability to keep records or organizations in the form of data sets.

6.4.1.5 Fog environment issues of load balancing and delay-aware scheduling

The strategies for load balancing, task scheduling, and performance measurements are the main topics of this section. When scheduling large-scale applications, the "cost-make span aware scheduling heuristic" is a strategy that seeks to achieve the balance between the use of cloud resources in the execution of the application and working cost, which is fixed as anticipated in this thesis, but it is inadequate for systems that must be run on a countless scale. Both fog computing and cloud computing use time more efficiently, which increases time consumption. We outline the potential issues using performance-based job scheduling and load-balancing strategies in this part. First, a description of the scheduling time for user task execution is given. For scheduling large-scale applications in such a setting, the trade-off problem is between the makespan and cloud cost. As a result, this research proposed a cost makes pan-aware scheduling heuristic scheduling algorithm. The main goal is to balance the performance of mandatory costs and application execution for the use of cloud resources, but this approach is unsuitable for large-scale applications (smart grid) due to the enormous volume and steadily rising number of service requests, as well as the time requirements of both cloud and fog computing platforms. It is suggested to use DEBTS (delay energy balanced task scheduling) for overall energy consumption, which reduces service delay and jitter.

Locations of fog nodes, connections to fog nodes that are available, traffic demands, and channel conditions are initially modified throughout time. The centralized controller uses a lot of energy, and if it fails, the entire network will collapse; a number of jobs take a long time to do. The response time is slow when there are more jobs to do. The bees life algorithm (BLA) finds the best compromise between the amount of memory allotted and the CPU execution time for offering fog computing services to mobile consumers. This is because it does not take into account how frequently new requests come in, while existing ones are still being processed for the fog computing environment. Fourth, basic fog architecture results from large delays at specific fog nodes for task execution or with regard to the number of tasks that have arrived. For fog clusters and nodes, an enhanced version of the non-nondominated sorting genetic algorithm II (NSGA-II) is suggested.

6.5 PROPOSED MODEL IN SCHEDULING

In this section, proposed fog computing architecture for task scheduling and load-balancing techniques fog architecture is classified as four-tier fog and is designed for delay-aware task scheduling and load-balancing approaches. In the tier–1 approach, IoT devices are used, and the process of data acquisition will be performed during this phase. Then comes the 2-tier approach that can connect two networks using routers, and the main task is to perform application processing with high-priority and low-priority codes. Here, a dual fuzzy logic algorithm has been performed for computations. In 3–tier architecture, purely fog nodes will be activated with clustering, scheduling, and load-balancing approaches of current fog nodes calculated. Here, K-means clustering algorithm and earliest deadline

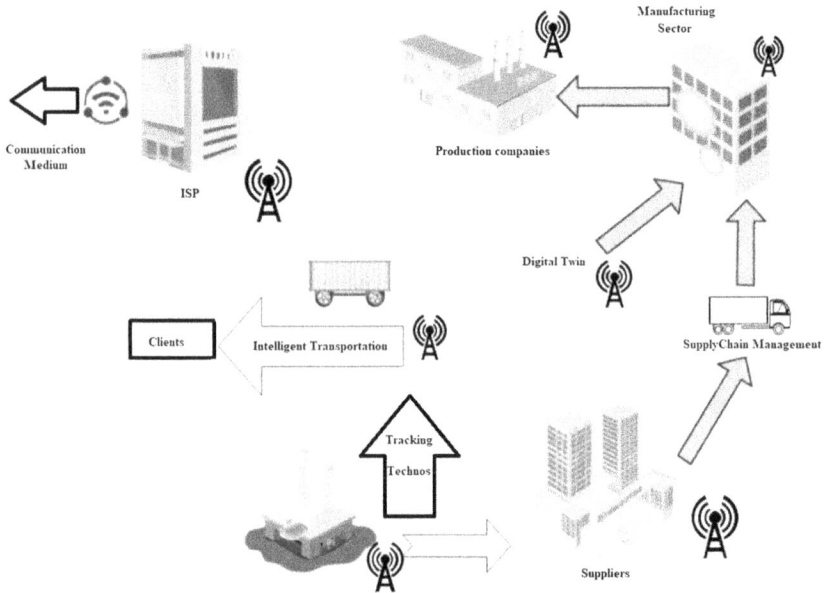

Figure 6.4 Proposed system for scheduling techniques.

first scheduling algorithms were used for the task-based process. A few parts of the ANN methodology also exist to calculate current nodes. ANN (artificial neural networks) focuses on convolution approaches to predict loading. In tier–4, cloud-based schemes were used, and we minimized the delay that was in the fog nodes. The earliest deadline first scheduling algorithm is performed (Figure 6.4). Figure 6.4 illustrates the proposed model for scheduling service.

Using the K-means++ clustering technique, fog node (FN) clustering has been created with 64 fog nodes because eight series of FNs have been taken into account. These are uses of the K-means ++ clustering algorithm that are clustered. Eight fractals are generated with 64 FNs, since each fractal has eight nodes. The IoT gadget transmits a request to the nearby FN. For this, every FN capacity and the current consumption are updated on a time-based basis. Each fractal selects a one-seed node to enhance the message. The arrangement is created by assembling nearby FNs into a group. Each fractal selects a one-seed node to enhance the message. Adjacent FNs are put together into a fractal for the layout, which groups the FNs into groups. Because the FN continuously communicates with contiguous FN with less energy consumption and higher bandwidth, one can easily customize the fractal scale, and this framework makes it easier to inform the fog fractals. The "Euclidean-distance" method is used to calculate the FN distance. This article suggests using K-means++ to choose the cluster center novel. Supposing, the shortest path from a point to the closest center, X.

6.6 CONCLUSION

Cloud computing is now the most cutting-edge enterprise technology in terms of resource availability, accessibility, scalability, and pricing due to its recent quick growth. Here, there is a significant rise in the majority of cloud components' security risks. This study focuses on the cloud infrastructure layer, where virtual machines and their counterparts are attacked. Virtual machines are at risk from inter-virtual machine (VM) communication assault, which allows co-located virtual computers to launch attacks against other virtual machines. A prominent phenomenon in the IT business is cloud computing. Cloud services are essential to millions of enterprises to achieve their goals. Researchers are attempting to put into practice an effective system for optimized resource allocation to lessen their needs. Therefore, this research aims to suggest an efficient resource controller based on prognosis that is machine learning based. The model in this instance is concentrated on two goals. The first and most important step is to forecast the resources using the best ML, which is compiled into the fog computing methodology, to allocate network resources. In this instance, we used fog computing to identify the need to build fog nodes and to design and implement a load-balancing method for the fog computing environment. The type of design chosen and the efficacy of the optimization algorithm employed to manage traffic load determine how effective fog networks are. The various steps of provisioning and its computing enable the thesis goals will be met. Autonomic computing should have the following fundamental qualities, such as being fully aware of itself and its parts, being able to adapt to unusual circumstances by changing how it is configured, always being able to optimize itself, being able to survive and recover from serious errors, and being able to defend itself against attacks coming from all directions, it must be heterogeneous and disguise from users the complexity of how it works, it must be aware of its working environment, and behave properly.

REFERENCES

Ajay, X., Sam Paul, P., Shylu, D. S., & Paul, M. P. (2021). Iot Based Prognostics Using MEMS Sensor with Single Board Computers for Rotary Machines. *Przegląd Elektrotechniczny*, 1(11), 172–176. 10.15199/48.2021.11.31

Baccarelli, E., Naranjo, P. G. V., Scarpiniti, M., Shojafar, M., & Abawajy, J. H. (2017). Fog of Everything: Energy-Efficient Networked Computing Architectures, Research Challenges, and a Case Study. *IEEE Access*, 5, 9882–9910.

Chitra Devi, D. & Rhymend Uthariaraj, V. (2015). Load Balancing in Cloud Computing Environment Using Improved Weighted Round Robin Algorithm for Non-preemptive Dependent Tasks. *Hindawi Publishing Corporation, The Scientific World Journal*, 2016, 1–14.

Dabbagh, M., Hamdaoui, B., Guizani, M., & Rayes, A. (2015). Toward Energy-Efficient Cloud Computing: Prediction, Consolidation, and Overcommitment. *IEEE Network*, 29(2), 56–61.

Dastjerdi, A. V., Gupta, H., Calheiros, R. N., Ghosh, S. K., & Buyya, R. (2016, January 1). *Chapter 4 - Fog Computing: Principles, Architectures, and Applications* (R. Buyya & A. Vahid Dastjerdi, Eds.). ScienceDirect; Morgan Kaufmann. https://www.sciencedirect.com/science/article/pii/ B9780128053959000046

Hassan, M. M., Abawajy, J., Chen, M., Qiu, M., & Chen, S. (2019). Special Section on Cloud-of-Things and Edge Computing: Recent Advances and Future Trends. *Journal of Parallel and Distributed Computing*, *133*, 170–173. 10.1016/j.jpdc.2019.07.004

Hussain, A., Aleem, M., Iqbal, M. A., & Islam, M. A. (2019). Investigation of Cloud Scheduling Algorithms for Resource Utilization Using Cloudsim. *Computing and Informatics*, *38*(3), 525–554. 10.31577/cai_2019_3_525

Jingle, D. J., & Paul, M. (2021). A Collaborative Defense Protocol against Collaborative Attacks in Wireless Mesh Networks. *International Journal of Enterprise Network Management*, *12*(3/4), 1. 10.1504/ijenm.2021.10030549

Khan, Md. A. M., Khan, M. R. J., Tooshil, A., Sikder, N., Mahmud, M. A. P., Kouzani, A. Z., & Nahid, A. A. (2020). A Systematic Review on Reinforcement Learning-Based Robotics within the Last Decade. *IEEE Access*, *8*, 176598–176623. 10.1109/access.2020.3027152

Khosravi, A., & Buyya, R. (2018). Energy and Carbon Footprint-Aware Management of Geo-Distributed Cloud Data Centers: A Taxonomy, State of the Art, and Future Directions. Sustainable Development: Concepts, Methodologies, Tools, and Applications. https://www.igi-global.com/chapter/energy-and-carbon-footprint-aware-management-of-geo-distributed-cloud-data-centers/189954

Kumar, M., Kumar Verma, H., & Sikka, G. (2020). A Secure Data Transmission Protocol for Cloud-Assisted Edge-Internet of Things Environment. *Transactions on Emerging Telecommunications Technologies*, *31*(6). 10.1002/ett.3883

Lin, C. (2012). A Novel Green Cloud Computing Framework for Improving System Efficiency. *Physics Procedia*, *24*, 2326–2333.

Nikravesh, A. Y., Ajila, S. A., & Lung, C. H. (2017). An Autonomic Prediction Suite for Cloud Resource Provisioning. *Journal of Cloud Computing*, *6*(1). 10.1186/s13677-017-0073-4

Paul, M., & Ravi, R. (2018). Cooperative Vector Based Reactive System for Protecting Email against Spammers in Wireless Networks. *Journal of Electrical Engineering*, *18*(4). http://www.jee.ro/index.php/jee/article/view/WG1519034616W5a8aa 0f8d0c62

Qu, G., Rawashdeh, O. A., & Hariri, S. (2009). Self-Protection against Attacks in an Autonomic Computing Environment. *22nd International Conference on Computer Applications in Industry and Engineering 2009, CAINE 2009*, 13–18. https://experts.arizona.edu/en/publications/self-protection-against-attacks-in-an-autonomic-computing-environ

Sharma, A., & Awasthi, I. (2018). Novel Technique for Load Balancing in Cloud Computing, *International Journal of Computer Applications*, *182*(4), 37–40.

Singh, A., Juneja, D., & Malhotra, M. (2015). Autonomous Agent Based Load Balancing Algorithm in Cloud Computing. *Procedia Computer Science*, *45*, 832–841. 10.1016/j.procs.2015.03.168

Younge, A. J., von Laszewski, G., Wang, L., Lopez-Alarcon, S., & Carithers, W. (2010, August 1). *Efficient Resource Management for Cloud Computing Environments*. IEEE Xplore. 10.1109/GREENCOMP.2010.5598294

Chapter 7

6G vision on edge artificial intelligence

B. Nivetha[1], Poongundran Selvaprabhu[1], Vivek Menon U.[1], Vetriveeran Rajamani[1], and Sunil Chinnadurai[2]

[1]Department of Communication Engineering, School of Electronics Engineering (SENSE), VIT, Vellore, Tamil Nadu, India
[2]Department of ECE, SRM University, Andhra Pradhesh, India

7.1 INTRODUCTION

Edge intelligence enabled by artificial intelligence (AI) technology is among the crucial element missing in 5G networks and is most likely to be present in the futuristic next-generation (6G) networks. 6G networks will be characterized by ubiquitous AI, featuring hyper-flexible architectures that provide human-like intelligence presented in Zhou et al. (2019). The restricted resource the absence of supreme quality data with labels, and the lack of optimized AI architecture are the three major challenges to the widespread adoption of AI in wireless networks. The author in Kalapothas, Flamis, & Kitsos (2022) provides a conceivable remedy to the aforementioned challenges is edge intelligence, commonly known as edge-native AI. Edge intelligence (EI) enables accelerated AI integration within the system in 6G wireless communication, which deploys a massive scale of edge servers (ESs) to process and make decisions based on AI that is closer to the data source of the service requests mentioned in Lin & Shen (2017). Additionally, the edge environment in Debauche et al. (2021) presents a special setting for deploying the applications based on processing units with various levels of intelligence capabilities due to its opportunistic nature and interaction with several stakeholders. Edge applications, which are reliant on the user-defined device and the edge infrastructure components are undergoing a paradigm shift driven by intelligence, context awareness, and autonomy. Using edge computing, computational units can be designed and distributed across layers of the system architecture that can also be decentralized vertically and horizontally are present in Liu et al. (2019). Computational units in modern AI can only support decentralization and distribution. AI and edge computing differ fundamentally in their premises, which makes challenging by bringing them together. An EI system, which is based on AI techniques such as machine learning (ML) and deep neural networks is already being regarded as a key element in 5G. As a consequence, performance enhancement, traffic optimization, and applications including ultra-low

DOI: 10.1201/9781003369028-7

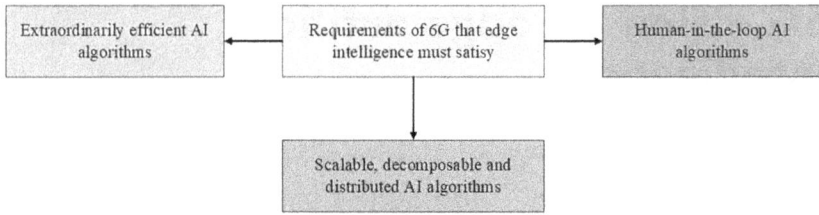

Figure 7.1 Requirements of edge intelligence in 6G.

latencies have benefited significantly from edge intelligence in 6G. Furthermore, leveraging edge intelligence will enable an entirely new class of products and services mentioned in Legaard et al. (2022). The EI that is required for 6G must provide extraordinarily efficient AI algorithms; intelligent human-in-the-loop AI algorithms; and scalable, decomposable, and distributed AI algorithms (Xiao et al., 2020). An extraordinarily efficient algorithm is both resource-efficient and data-efficient. Figure 7.1 illustrates the significant requirements of 6G that edge intelligence must satisfy.

The main contribution of this chapter is to highlight the fundamental needs and developments that shape the EI in 6G that is particularly from the standpoint of artificial intelligence. Table 7.1 follows a few interesting concepts related to 6G edge intelligence. The structural organization of this chapter is depicted in Table 7.2.

7.1.1 Emerging technologies in 6G wireless network

Depending on the previous development of wireless networks, 6G networks initially rely mostly on the 5G architecture and retain the advantages of 5G (Tamilarasan, Selvaprabhu, & Venkatesan, 2023). Due to this development, there will be some new technologies are included and get enhanced for 6G. The 6G environment system can be activated by a variety of technologies mentioned in Jia (2013). A few key potentials of 6G technologies are covered below.

7.1.1.1 Optical-free technology

Radio frequency-based communications for all conceivable devices gets network access and optical free technique to anticipate the 6G communications. These connections can interface the network to backhaul/front-haul network access. Since 4G communication systems, these technologies have been in use. The performance of these technologies has been improved, and problems with them have been resolved. It can achieve high data rates and low latencies with secure communications.

Table 7.1 A few interesting concepts related to 6G edge intelligence

Reference	Year	Summary
(Yang, Qin, Cheng, Xu, & Zhao, 2022)	2022	First, the authors in this manuscript provide a systematic overview of AI towards 6G. Following this, a detailed illustration was given of the requisites, issues, and current trends of distributed edge intelligence in the upcoming 6G networks. In addition, the 6G networks that play a significant part in both industry and academia in support of the versatile and liquid-specific applications are also discussed.
(Dong, Wu, Zhang, & Ding, 2022)	2022	This manuscript investigates the notion of edge intelligence from the semantic cognitive viewpoint. Additionally, the authors discussed and envisioned edge semantic cognitive intelligence in the context of conceptual models, frameworks, uses, and future research directions.
(Gong, Yao, Wang, Li, & Guo, 2022)	2022	Taking into account a wide range of industrial internet of things (IIoT) tasks, the authors in this manuscript propose an innovative IIoT prototype with edge intelligence functionality in 6G. Furthermore, to cooperatively optimize the difficulties of the IIoT system with the distribution of transmission resources and unloading devices, the authors also developed a unique deep reinforcement-based network topology, followed by an adaptive intervention clustering and the isotone action generating method.
(Wang et al., 2022)	2022	The VANET, a potential contender for the future 6G networks is gaining significant attention due to the hike in the number of smart cars. Although the use of blockchain are existing, it is not as efficient as the proposed scheme. In the proposed scheme, the authors present an effective data-sharing scheme for protecting privacy in the 6G VANET based on the blockchain and edge intelligence.
(Zhu et al., 2023)	2022	With the advent of over-the-air federated edge learning (FEEL), edge AI beyond 5G and 6G networks is now possible. To improve the efficiency of communication for FEEL towards edge intelligence, the authors in this manuscript provide a systematic review of air FEEL, which applies cutting-edge over-the-air data aggregation or over-the-air computation techniques. In addition, fundamental ideas, difficulties with the design, cutting-edge methods, and numerous research problems with air FEEL are also discussed.
(Chang et al., 2022)	2022	In this manuscript, the authors provide a comprehensive survey of the unification of 6G-enabled edge AI and the metaverse. In addition, the authors also explored the novel form of metaverse architecture that utilized 6G-enabled edge AI, some of the critical problems, unresolved issues, and potential implications for research related to 6G-enabled edge AI for the metaverse.
(Tang et al., 2021)	2021	Using a deep reinforcement learning technique for networks beyond 5G or 6G, the authors of this manuscript examined how to optimize FEEL in unmanned aerial vehicles with limited battery life.
(Gupta, Reebadiya, & Tanwar, 2021)	2021	Having extremely reliable low-latency applications where a delay of a millisecond is unbearable, the authors of this manuscript have proposed an edge intelligence system based on 6G. Drone web, wearable technology, border surveillance, driverless vehicles, and telesurgery are a few of the applications that are specifically discussed in this manuscript.

Table 7.2 The structure of this chapter associated with edge AI in 6G

Sections	Subsections	Perspectives
7.1 Introduction	7.1.1 Emerging technologies in 6G wireless network	7.1.1.1 Optical-free technology 7.1.1.2 Quantum technology 7.1.1.3 Native network slicing 7.1.1.4 Integrated access backhaul networks 7.1.1.5 Holographic beam forming
7.2 Effective Training Models	7.2.1 Edge AI learning models 7.2.2 Wireless technique edge training	7.2.1.1 Federated learning 7.2.1.2 Decentralized learning 7.2.1.3 Split learning 7.2.1.4 Distributed reinforcement learning 7.2.1.5 Trustworthy learning 7.2.2.1 Over-the-air computation 7.2.2.2 Massive MIMO 7.2.2.3 Reconfigurable intelligence surface
7.3 Effective Edge Inference	7.3.1 Horizontal edge inference 7.3.2 Vertical edge inference	7.3.1.1 ED distributed inference 7.3.1.2 ES cooperative inference 7.3.2.1 ED-ES co-inference 7.3.2.2 Low latency and ultra-reliable communication 7.3.2.3 Task-oriented communication
7.4 Architecture for Edge AI in 6G Wireless Network	7.4.1 Centralized architecture 7.4.2 Decentralized architecture 7.4.3 Hybrid architecture 7.4.4 Self-learning architecture 7.4.5 End-to-end architecture 7.4.6 Data governance	7.4.6.1 Independent data plane 7.4.6.2 Multiplayer and multi-domain roles 7.4.6.3 Management and orchestration of edge AI
7.5 Edge AI Application towards 6G	7.5.1 Characteristics of metaverse 7.5.2 Edge AI-based metaverse architecture	7.5.1.1 Immersive 7.5.1.2 Multi-technology 7.5.1.3 Interoperability 7.5.1.4 Sociability 7.5.1.5 Longevity 7.5.2.1 Edge cloud metaverse (ECM) architecture 7.5.2.2 Mobile ECM Architecture 7.5.2.3 Decentralized metaverse architecture
7.6 Challenges and Applications of Edge AI in 6G	7.6.1 Challenges in edge AI 7.6.2 Some more futuristic applications of edge AI in 6G	7.6.1.1 Adversarial learning and adaptation 7.6.1.2 Interpretable AI 7.6.1.3 Quality of experience 7.6.1.4 Interactive AI 7.6.1.5 Detecting and predicting human intention 7.6.1.6 Intelligent human to machine communication 7.6.2.1 Industrial Internet of Things 7.6.2.2 Healthcare 7.6.2.3 Autonomous driving Vehicles 7.6.2.4 Security and privacy 7.6.2.5 Education

7.1.1.2 Quantum technology

A 6G enabler is thought to be the newly developing concept of quantum computing and quantum ML with its interactions in communication networks. To overcome challenging problems, quantum computing needs to develop and solve complex issues. By using a quantum process by incorporating the quantum no-cloning hypothesis and the uncertainty principle, the author in Chen, Su, & Xu (2019) delivers outstanding security.

7.1.1.3 Native network slicing

A network connection can provide committed virtualization to assist the efficient delivery of any service to a variety of customers, cars, machines, and industries owing to native network slicing.

7.1.1.4 Integrated access backhaul networks

6G networks have attained a maximum density. Backhaul connection, such as fiber-optic networks, is related to each of the access networks. The majority of access networks are handled by tightly integrated access and backhaul networks.

7.1.1.5 Holographic beam forming

Through the use of signal processing, an antenna array is directed towards broadcast radio signals in a particular direction using beam forming. It is a component of enhanced antenna systems or adaptive antennas. High noise signal, interference avoidance, interference rejection, and excellent network are the few benefits of beam-forming technology. In contrast to MIMO systems, holographic beam forming (HBF) is a unique novel beam-forming technique that makes use of software-defined antennas presented in Jiang et al. (2019). HBF is a beneficial solution in 6G for the effective and adaptable transmitting and receiving signals in multiple-antenna communications systems.

7.2 EFFECTIVE TRAINING MODELS

7.2.1 Edge AI learning models

This chapter describes the various communication of optimized distribution algorithms for edge training before moving on to promising wireless strategies that assist the implementation of edge learning models. A failure or empirical hazard function obtained in the implementation of the learning process is often minimized. However, this model fits an optimization method using distributed data produced by a massive variety of intelligent systems.

7.2.1.1 Federated learning

This learning technique enables a significant number of decentralized devices associated with the various services which cooperatively train a common global mode (e.g., anomaly detection, remuneration committee, next-word prediction, etc.) by utilizing locally gathered data sets. In federated learning (FL), a cooperative ML framework develops the global predictive method without accessing the private original data of the edge device (ED). It can provide the specialized ES in charge of aggregating local and distributing global learning model upgrades. Cross-device FL in Liu et al. (2020) poses various challenges than cloud-based data distribution learning. This includes the high transmission costs with a massive model which can frequently be substituted over wireless connections. Using this learning model, multiple data can collaborate to train a significant deep learning model without revealing their data sets. Each client trains its distinct data by utilizing a local ML model. The learning models among all the clients are combined by a server at the completion of each training iteration. Federated averaging is an impactful approach to minimize the number of interaction rounds by carrying out multiple local upgrades such as managing multiple stochastic gradient descent (SGD) implementations on each ED. The author in Yang et al. (2020) offered numerous defensive techniques to address the challenges from the perspectives of an aggregation algorithm, a detection mechanism, and reputation management, as shown in Table 7.3.

The author in Duan et al. (2022) provides the most extensively used algorithm for the fundamental learning model is the federated average algorithm presented in Algorithm 1. The global model of this algorithm is subsequently initialized using the parameter M. Each time server chooses M out of a total of B clients to take part in the current round of training.

Table 7.3 Various defense solutions using the federated learning model

Aggregation algorithm	Detection mechanism	Reputation management
In the FL process, aggregation is a crucial activity that has a direct impact on the outcomes of model convergence. The goal of this algorithm is to significantly lessen the negative effects of low-quality model upgrades produced by malicious devices on overall model training.	Finding malicious devices and preventing them from getting involved in FL training is another obvious solution. Such a technique has often used the integrity of the sub-model produced among devices as an evaluation criterion to spot fraudulent equipment.	More dependable devices are indicated by a high reputation value by using a metric called reputation. The historical actions of the devices can be utilized as a significant signal to assess their dependability and trustworthiness.

However, the selected client with their respective global parameter M_t^G are supplied to the corresponding client is referred to as the client upgrade function. Here K, E, ω, t are the local batch size, local epochs, learning rate, and respective time slots. Meanwhile, U_t, F_B, l and A are the random database in client, dimensional of matrix B, length of data, and random batches. The participating client in this upgrade function will initialize the received parameter $M_{(t+1)}^C$ of the local model to train the local database. These upgrade function again once the completion of the local training database is done. The server gathers local parameters from all involved clients and aggregates them using a weighted median to upgrade the global model.

Algorithm 1 Federated Average Learning Model.

/* Server Execution */

1:   **Initialize** M, K, B, E, ω.
2:   **For** each round t = $1,2, \ldots .. M$ do
3:      $U_t \leftarrow$ Select random B clients.
4:      **For** each client $B \in U_t$ do
5:         $M_{(t+1)}^B \leftarrow$ Client update (B, M_t^G).
6:      **End For**
7:      $M_{(t+1)}^G \leftarrow \Sigma_{B=1}^B \frac{|F_B|}{|F|} M_{(t+1)}^B$.
8:   **End For**
9:   /* Client Execution*/
10: **Initialize** $M \leftarrow M_t$.
11: $A \leftarrow$ Split B into a batch of A.
12: **For** each E (i.e.,) from l to E do
13:     **For** batches a to A do
14:        $M \leftarrow M - \omega \nabla l(M; a)$.
15:     **End For**
16: **End For**
17: Return M to server.

7.2.1.2 Decentralized learning

Decentralized deep learning (DL) provides a wide range of potential for possible application uses in autonomous industrial facilities, such as the cooperative approach for mapping and localization, cooperative automatic driving, collaborative robots in high-tech manufacturing, and production environment. This learning architecture has specific benefits concern on data localization, computation scalable processing, and efficient communication are presented in Lee et al. (2020). Meanwhile, swarm learning generates a complete decentralized AI platform based upon the decentralized DL to construct and maintain the local data sets at each ED. This type of learning enables high levels of privacy, data protection, resiliency, and scalability. During this learning model, a multitude of decentralized device-to-device communication systems and their corresponding protocols including connectivity graphs can be supported instead of FL architecture. Moreover, it can fix the heterogeneous hardware straggler issues and make the system more resilient to data poisoning threats and network congestion. The converging characteristics of this learning model significantly rely on a decentralization averaging framework and structure of the data exchange network. Improved communication effectiveness for switching the locally upgraded models at the ED to its corresponding peers, reduces the transmission rounds (i.e., enhanced the rate of convergence) or the quality of information transmitted per round. The reduction of variance among a certain gradient tracking approach was specifically examined to attain a rapid convergence rate. The performance of the periodic decentralization including the execution of numerous local modifications systematically reduces the average number of device communication rounds. Moreover, the quality of the exchanged information can be minimized by quantizing the periodically generated models in the communication systems. A consensus line-of-sight regulating mechanism was developed to accomplish the trade-off for both efficient learning and the accuracy of periodically averaging the decentralized ML. Additionally, a network layout is essential to increase the effectiveness of communication. For this purpose, a group of switching direction techniques of multipliers was proposed that divides the employees into head and tail workforce to generate a connectivity chain. Consequently, a momentum-based approach has been established to produce strong generalization performance and to solve the heterogeneity problem of local data sets in this learning model. The models illustrating the notion of FL and decentralized learning are depicted in Figure 7.2.

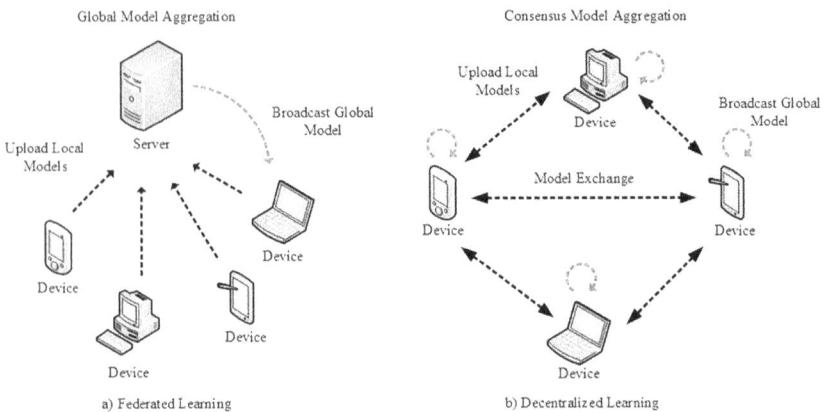

Figure 7.2 (a) Federated and (b) decentralized edge learning models.

7.2.1.3 Split learning

The split learning model supports a cooperative training process between the edge serves and devices by distributing the parameters among the edge nodes. The learning model is classified into two categories like edge client (EC) and ES model which is implemented in the network system. Most of the primary training databases are analyzed and stored on the EC rather than being transmitted towards the ES, which would be related to FL. The entire model is trained via a sequence of forward and backward propagation among the ES-EC network. The client employs the forward propagation till the cut layer (i.e., at the final layer of EC) while using the training database to load the EC model. The execution of the cut layer is subsequently sent to the ES as a "smashed database" that is usually together with the appropriate label. The ES begins the backward propagation process by considering gradients and upgrading the parameters of each layer of ES after estimating the loss function. The EC has then received the gradients function of the smashed database by the ES.

This propagation process is continued until the maximum reach of convergence rate at the complete network model. During a round-robin fashion, this learning framework with numerous EC accesses the data from several entities. Each EC interacts with the ES by using alternating iterations. Multi-EC in this learning requires the coordination of sub-models on various clients to maintain model consistency. As a consequence, each EC must upgrade its model parameters before beginning the subsequent training iterations. Both centralized and peer-to-peer connectivity techniques can be used to coordinate EC models. While training a centralized mode, an EC upgrades the parameter of the models to the ES where each subsequent EC in retrieves them. In peer-to-peer mode, the ES transmits the location of the last trained EC to the currently available training EC, which then links to that EC and retrieves the parameter of the model. Consequently, this framework is conceivable to train EC models without cooperation. However, there seems to be no guarantee that such asynchronous model training would lead to convergence. The model illustrating the notion of split learning and distributed reinforcement learning is depicted in Figure 7.3.

7.2.1.4 Distributed reinforcement learning

This learning model provides a technical approach for consecutive outcomes in versatile circumstances by direct interaction with a challenging environment. Model, policy (e.g., spontaneous policy gradient), value-based (e.g., Q learning), and actor-critic approaches are the common reinforcement learning techniques. Multi-agent reinforcement learning (MARL) defines multiple agents cooperatively communicating with a normal scenario to accomplish the desired outcome and maximize collaborative teamwork with

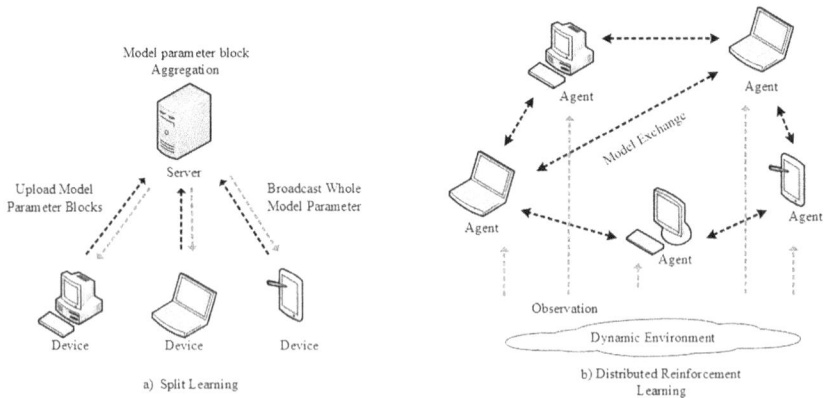

Figure 7.3 (a) Split and (b) distributed reinforcement edge learning models.

different local action spaces. Each agent can collaborate with other service stations on the wireless network to determine the most appropriate programming configurations to enhance the network's quality of service. An effective communication method between multiple agents should become crucial to maintaining steady performance for MARL to attain vast state-action space, delayed incentives, and varied agent behaviors.

7.2.1.5 Trustworthy learning

To develop and implement AI systems for high-stakes applications at the edge networks, it is essential to provide secrecy and safety, ease of understanding, obligation, and resilience for the edge learning models. Therefore, the heterogeneity of huge-scale edge systems and distributed data sets develops new challenges in the implementation of trustworthy edge AI algorithms. Introducing random perturbations, differential secrecy offers a lightweight confidentiality approach to ensure a threshold of privacy exposure for local data sets. In Li et al. (2021), the authors addressed wireless network's inherent additive noise and signal aggregation features can be exploited as a privacy-preserving technique. The generated intrinsic noisy model can be modelling the privacy exposure to the database at the ES without incurring learning performance. Edge AI must provide user privacy while being resilient to errors and hostile attackers because its decentralized approach makes it simple for external hackers to probably take over its learning process. Figure 7.4 elucidates the trustworthy edge learning model.

7.2.2 Wireless technique edge training

This chapter uses the configuration of edge AI modeling techniques to abide by the principles and design of wireless communication when the network

Secure Model Aggregation

Broadcast Global
Models

Server

Upload Encoded
Local Models

Devices   Malicious Devices   Colluding Devices

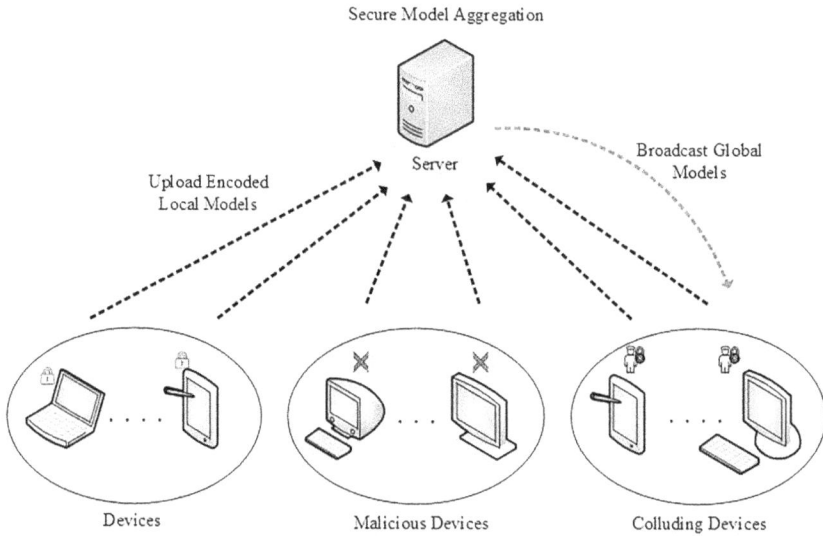

Figure 7.4 Trustworthy edge learning models.

purpose for edge AI reverts from the traditional data rates to the learning performance. The co-design principle for learning and communication for future 6G wireless technologies enables AI features that reside inherently within 6G. This clears up several misconceptions about edge training model efficiency in wireless networks. To support the enormous number of ED actively engaged in the training process, we provide more effective multiple antenna approaches and next-generation multiple access systems.

7.2.2.1 Over-the-air computation

To upgrade a global model, edge training tasks frequently include generating aggregation functions of numerous local model modifications presented in Weng & Qin (2021). The typical aggregation in FL, normative accumulation in decentralized learning, and resilient accumulation in trustworthy learning achieve the local upgrades from the ED and determine the proper aggregation function at the ES. Air computation offers a convenient multiple access method for a limited latency aggregation model. This utilizes interference to curtail networking bandwidth usage by simultaneously transmitting the locally upgraded models. The received signals at the ED are multiplied by the Rayleigh fading channel and then superimposed over the noise to deliver a noisy weight value of the broadcast signals. The delayed communication and required bandwidth of air computation do not increase the number of ED. Meanwhile, it eliminated the hurdle in the edge learning process.

7.2.2.2 Massive MIMO

Massive antenna arrays are a crucial component of wireless networks that helps to obtain significant energy and spectral efficiency in 6G. Performance over a wider frequency range and current advancements in digital, analog, and hybrid beam forming have aided in the implementation of massive MIMO (Menon & Selvaprabhu, 2022). It has been established to have considerable benefits for edge training systems, such as high model accumulation rates and accuracy along with high reliability for enormous device connectivity. Utilizing highly scalable ED, ultra-dense wireless networks have become an effective way to ensure great reliability and minimal delay and performance edge training over a large geographic region. The straggler problem (i.e., devices with limited computational capabilities may extend the training time) and undesirable channel characteristics are negated by concurrently transferring massive local model upgrades with heterogeneous virtual ES with widely available communication. Furthermore, in contrast to the single ES model, distributed ES is tolerant to host the interruption issues for sustained edge training. Figure 7.5 gives a detailed illustration of over-the-air computation and massive MIMO wireless techniques for edge training model.

7.2.2.3 Reconfigurable intelligence surface

Generally, in Yuan et al. (2021), magnitude synchronization by modulating the transmit signals of reconfiguration intelligence surface (RIS) is necessary

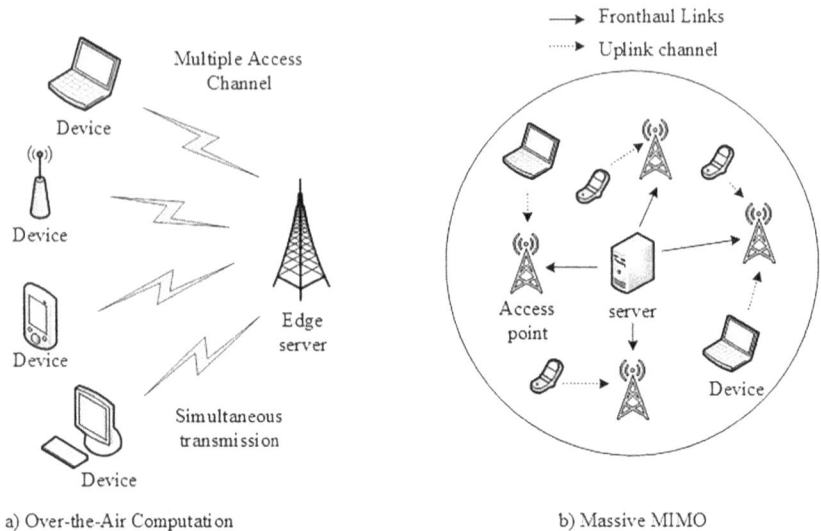

a) Over-the-Air Computation

b) Massive MIMO

Figure 7.5 (a) Over-the-air computation and (b) massive MIMO wireless techniques for edge training model.

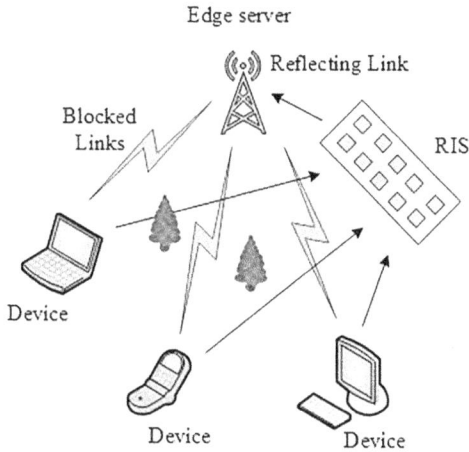

Figure 7.6 Reconfigurable intelligent surface wireless techniques for edge training model.

to acquire the intended average process of local model upgrades for model accumulation via Air Comp. However, because of the resource-constrained ED and the uneven fading channels, the undesirable multipath propagation architecture invariably causes magnitude reduction and lack of alignment with the nonlinear model accumulation, which reduces the training capacity of the edge learning process.

In a nutshell, it is possible in RIS that the enormous ED are placed in a network dead zone with erratic access to the ES which makes it difficult for inadequate channel links to detect device activities. A major challenge to the implementation of edge AI systems is the heterogeneity of processing, connectivity and retention across ED. By establishing advantageous propagation linkages via RIS, it is possible to significantly shorten the latencies for local model upgrades of the active devices while reducing stragglers. Figure 7.6 demonstrates the RIS wireless techniques used for edge training model.

7.3 EFFECTIVE EDGE INFERENCE

This chapter implies categorizing edge inference into two categories based on various processing collaboration schemes: horizontal edge inference (i.e., data processing resources can always be obtained between many ED, or just be aggregated among ES) and vertical edge inference (i.e., data processing resources can be produced among ED and ES), which are respectively addressed in the next two subsections.

7.3.1 Horizontal edge inference

In premise, this chapter discussed the interference coordinating methods for task-oriented low latency concepts, horizontal, and vertical edge inference, respectively.

7.3.1.1 ED distributed inference

Massive efforts have been made on ML with DL design compression and neural network design, which are explored to enable energy efficiency and low-latency model inference among single devices with constrained storage and processing resources. Due to the constraints of storage capacities at ED, it becomes incredibly challenging for applications like mobile navigation. This contains a significant map information database to complete inference operations on a single device. While consulting the cloud data hub, ED-based wireless distributed inference provides lower latency, high (accuracy, scalability, and robust services) for the ED presented in Li et al. (2019). The ED distributed inference particularly implies possible processing values depending on the input database utilizing the map function. Then it can share the possible values via effective horizontal communication between the ED to construct the preferred information processing or inference results while using an eliminate function. A dual uplink and downlink design focusing on the interference alignment notion was devised, which boosts the data rates for intermediated values rearranging rather than simply lowering the number of communication bits to further increase spectral efficiency.

7.3.1.2 ES cooperative inference

A promising procedure requires deploying and running DL models on ES. However, the primary constraint for ES cooperative inference acts as a constrained wireless bandwidth among the ED and ES. To diminish the uplink communication operating costs, it is imperative to compress and encode the input data source at the ED. Diverse data dimensional reduction algorithms have been proposed for this purpose by utilizing unique compute workloads and communication settings. Additionally, it is crucial to design extremely effective downlink services for bringing the output inference performance for the ED for applications requiring high production inference results. It has been observed that computation replication works well in reducing communication delay during computation offloading whenever the output size is high. To achieve this, each inference work is executed at several ES and indeed the inference results for the number of ED are immediately transferred through downlink cooperative transmission. While ES cooperative inference through downlink transmission collaboration can considerably increase network efficiency by reducing interference and channel uncertainty, it uses more energy to run identical inference tasks at different ES.

7.3.2 Vertical edge inference

Let us explore two different scenarios for vertical edge inference, one with a single ED and the other with numerous ED. Before introducing task-oriented communication, a unique design approach is recommended for vertical edge inference presented in Foukalas & Tziouvaras (2021). The practical solutions for these two efficient vertical edge inference scenarios are discussed.

7.3.2.1 ED-ES co-inference

Despite having low latency in ED distributed inference, its accuracy is constrained by its computing power and bandwidth. ES cooperative inference can produce good results with DL models that are precise. However, there is a risk of data leaking and excessive communication latency. ED-ES co-inference is an alternative to horizontal edge inference that promises to reduce communication overhead expenses while attaining high accuracy and privacy for deriving deep neural network models by empowering ubiquity AI services over a variety of application scenarios. Figure 7.7 illustrates the ED distributed and ES cooperative horizontal edge inference. Split model selection for the neural network is crucial to achieving the best information processing trade-off towards vertical edge

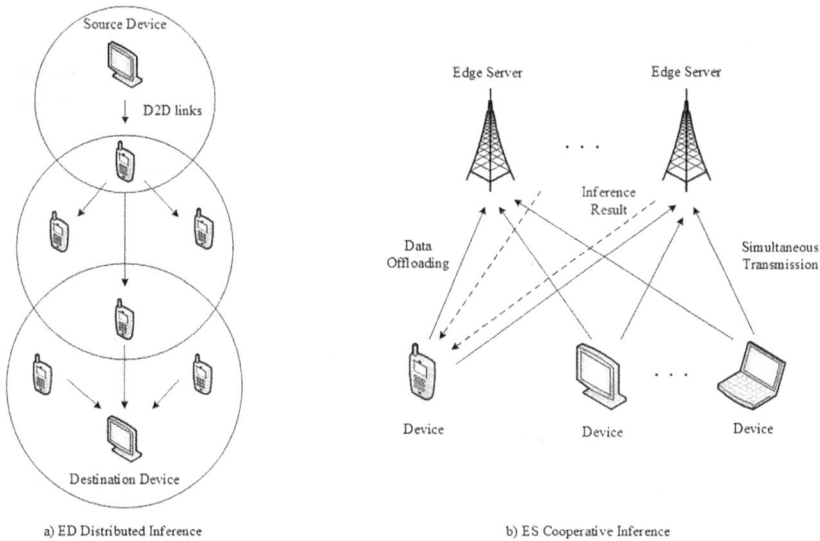

a) ED Distributed Inference   b) ES Cooperative Inference

Figure 7.7 (a) Edge device distributed and (b) edge server cooperative horizontal edge inference.

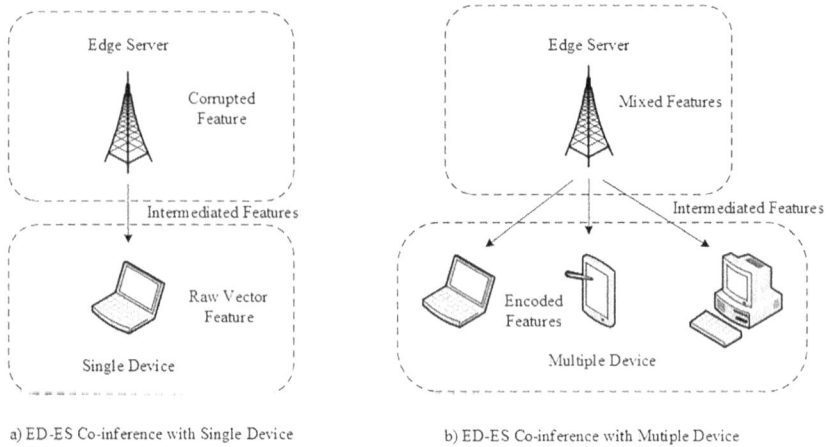

a) ED-ES Co-inference with Single Device   b) ED-ES Co-inference with Mutiple Device

Figure 7.8 Edge device–edge server co-inference for both (a) single and (b) multiple devices.

inference system through ED-ES cooperation by adaptively discretizing the computation responsibilities between ED and ES partnership efforts. Real-time vertical edge inference implementation has specific challenges because of the quick message transmission and signal amplification effect of the on-device split models extracted feature. The ED-ES co-inference for both single and multiple devices is also illustrated in Figure 7.8.

7.3.2.2 Low latency and ultra-reliable communication

The retrieved output features packet length might be very small, which lowers the possible data rate in this limited block size regime and increases the likelihood of a non-vanishing decoding error. In mission-critical programs, the outcomes of production inference from the ES should be transmitted to the ED with delay and reliability assurances. Cross-layer optimization is required to reduce the latency for ED-ES co-inference approach that considers the system dynamics, especially task characteristics throughout the network layer and wireless channel dimensions in the physical layer. Massive MIMO can be used to provide a practically deterministic networking environment and allow numerous ED to upgrade median features via short packet transfer while combatting channel rapid fading.

7.3.2.3 Task-oriented communication

Conversely, designers should construct the broadcast communication scheme in a task-oriented manner, i.e., just by delivering the information that is essential to the ES downlink inference task. When accessing the communications network design, the task-oriented communication concept represents a significant transformation from recovering data to task completion. It was initially evaluated in vertical edge inference using end-to-end learning and combined channel encryption that reduces the amount of on-device communication and computation overhead (Liu et al., 2019). The received signal, which has been tampered with by noise and channel fading, is processed immediately at the ES to produce the inference findings rather than having to decode the intermediate features. Providing robust, low delay, and power-efficient edge AI inference services will estimate and stimulates the co-design of wireless network systems and DL models.

7.4 ARCHITECTURE FOR EDGE AI IN 6G WIRELESS NETWORK

In terms of network infrastructure, network (function and administration), data governance, operations, and applications, the author provides a centralized, decentralized, hybrid, and end-to-end architecture design.

7.4.1 Centralized architecture

Through this architecture, the collection of data and model (training, inference) phases of AI model training are handled by the cloud server. Particularly, peripheral devices tried to implement at the edge layer like smart watches, cellular mobile, automobiles, and security cameras produce and gather data crucial for AI learning model. The server then utilizes the information transmitted by these ED to train AI models, which are then included for model inference.

7.4.2 Decentralized architecture

Under this concept, every edge computing node completes the cycle of AI learning model. Each edge interface device in (Rasheed, San, & Kvamsdal, 2020) specifically uses its local data to calibrate the AI model. Through communication networks, these edge processing nodes subsequently communicate model data. By exchanging local model upgrades, these nodes eventually acquire a globally integrated AI model. Without the assistance of cloud servers, edge computing nodes in this architecture can train AI models.

7.4.3 Hybrid architecture

A fusion of both centralized and decentralized architectural functions is referred to as hybrid architecture. The edge terminus devices are in charge of the local training phase and exchanging local model upgrades, while the ES serves as the architecture's central focus. More specifically, the ES is in charge of streamlining the overall AI model and distributing the upgraded AI model to ED over the network connection. To achieve the benefits of ES computing and resource efficiencies, the data set from the edge nodes can also be transferred at the same time to the ES for coordinated training.

7.4.4 Self-learning architecture

This edge AI architecture has the potential to drastically minimize the human involvement in data handling and forecasting models by generating the self-detection and modification over unidentified events, and most interestingly, automatic development, and evolution in response to changes in observed data. We present a self-learning architecture that relies on self-supervised generative adversarial networks (GAN) with numerous generators. Further with aim of this learning characteristics and building ML models to recognize and categorize unfamiliar new services from unprocessed sharing data dispersed across a vast geographical area. To produce random numbers that can accurately represent the various congestion data produced from numerous services across various network locations coverage, the proposed framework uses the generative learning characteristics of the GAN approach, which involves training multiple generators. This demonstrates the feasibility to train each system to generate simulated data samples that follow a relatively similar distribution as the actual traffic information combined with each group by introducing a classification model to maximize the distribution discrepancy of simulated data generated by different generators. These generators and classification techniques of the suggested architecture can be distributed individually across many ES, each training the model with a particular set of data, therefore reducing the computational strain on the ES.

7.4.5 End-to-end architecture

New services and their functionalities had been enabled at the design level for every successive generation of wireless networks, to accomplish escalating and frequently more demanding needs. The mobile network was initially designed and supposed to be used to provide voice services, since voice and packet traffic on mobile have been considered a centralized and hierarchical organization that is reflected in the architecture and implementation of networks. The ambition of "connected intelligence" requires the deep integration of interaction, AI, computing, and sensing at the ED. To achieve this, 6G will shatter and transition this traditional concept in favor of a unique design and architecture that meets these new requirements and is therefore enabled by both evolutionary and revolutionary empowering technologies. This provides a comprehensive end-to-end architecture for scalability and reliable 6G edge AI systems through this revolutionary design philosophy. The proposed architecture grants a

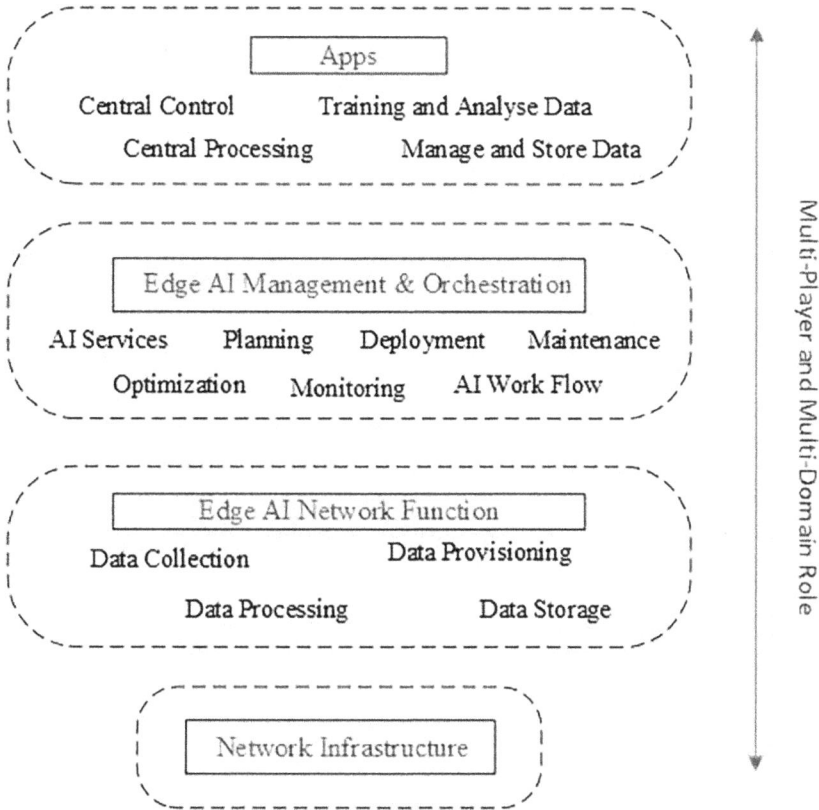

Figure 7.9 End-to-end architecture for ddge AI systems.

flexible and adaptable platform to support a variety of edge AI technologies with heterogeneous services provided by offering new wireless network facilities that enables effective data governance, trying to integrate communication and estimation at the edge, together with conducting automatically and configurable edge AI executives and orchestration. The end-to-end architecture for edge AI systems is illustrated in Figure 7.9.

7.4.6 Data governance

An enormous energy usage anticipated by considering security and confidentiality concern, we thought data in developing 6G networks will require data to be gathered, analyzed, and retained at the ED networks. There should be an endrosed for a uniform and effective data governance system at the architectural level because data and AI technologies in 6G are anticipated to be significantly more diverse than in previous generations. It reaches well beyond standard data collecting and archiving, which includes considerations for data (quality, availability, and sovereignty), along with knowledge creation and legal implications. Meanwhile, it needs to examine the procedure that adheres to regional or national information coverage requirements of the data provider when addressing the usage of obligations and rights.

7.4.6.1 Independent data plane

5G established a new data transmission analytics function (NWDAF) in the central network to deploy AI-based techniques for control, optimizing network-related functions such as AI-based scalability and enhancing client service satisfaction among other things. One of the major purposes is to gather information from other 5G network components to train AI learning models while using AI deduction for controlled and flexible system optimization. Additionally, 5G radio access networks adopted comparable processes, such as gathering and evaluating databases based on the existing self-organizing system and reduction of drive tests. A more cohesive and efficient paradigm for 6G needs to be developed from such a disparate approach to data collection and analytics.

In 6G, an autonomous data plane could aid in effectively organizing and managing data together while bringing privacy protection into account. Utilizing multi-domain data paves the road for organically integrating edge AI into 6G networks.

7.4.6.2 Multiplayer and multi-domain roles

The data governance environment contains a variety of roles, including data (controller, steward, provider, and supplier). These can be utilized by several business entities including individual users or by the same one. It becomes vital to design a multi-party information trading system to negotiate information rights and price-diverse corporate entities while assuring honesty, transparency, and efficiency. It has been done by utilizing decentralized technology including a blockchain with a digital ledger design. In Selvaprabhu et al. (2022), the authors enhance the business environment and data efficiency for the implementation of edge AI.

7.4.6.3 Management and orchestration of edge AI

Edge AI entails a complex partnership for communication, processing, and intelligence along with a wide range of learning models and algorithms, technology deployment, and other elements. Thus, establishing a paradigm for edge AI control and orchestration becomes crucial for the architecture-level design of native AI support. This paradigm must be developed to make it easier for the deployment and seamless integration of AI solutions, particularly those provided by third parties. To accomplish this, one can plan, install, manage and optimize the decentralized ML models, methodologies, edge technology deployment, and capabilities. The administration and orchestration of edge AI also should consider the AI process, global data, method of investigating along with diverse network resources, etc. Such a system will face significant scaling and cross-domain hurdles, which may involve challenging standardization initiatives. Therefore, it is not possible to design such a conceptual platform that will entirely rely on standards.

7.5 EDGE AI APPLICATION TOWARDS 6G

The metaverse is undoubtedly one of the most significant and exciting applications of the forthcoming 6G wireless communications technology. This application seeks to establish a society where virtually and reality coexist that can support thousands of millions of internet users. This indicates that the metaverse imposes more requirements on the current existence of edge AI architecture.

7.5.1 Characteristics of metaverse

The metaverse users in Araujo Inastrilla (2023) can simulate their experiences as those of people in the real world by interacting with customized three-dimensional (3D) virtual representations. For instance, virtual world series avatars can engage in a wide range of human activities like virtual meetings, freelancing, and shopping. The metaverse contains the following characteristics specifically.

7.5.1.1 Immersive

For the past several years, screens and mobile devices have found it challenging for users to engage with virtual worlds. However, the blending of the real and virtual worlds has greatly lessened this restriction whereas for the ongoing development of metaverse technology. The "stimulus" an avatar receives in this application can be highly realistically returned to the user in the real world owing to extended reality (XR) technologies and smart wearables like brain-computer interfaces (BCI). For instance, users can access the metaverse for distant video conferences while wearing virtual reality (VR) spectacles and augmented reality (AR) headphones, realizing the transition from the two-dimensional (2D) interactive platform of the physical universe to the 3D simulation domain of the virtual environment. This could make the user feel increasingly immersed.

7.5.1.2 Multi-technology

Digital twins, XR, blockchain technology, object recognition, and cutting-edge digital and intelligent systems are present in the metaverse. Particularly with twin technology, users can reconstruct and test physical parameters in the metaverse by producing multi-scale, high-fidelity real-time maps of those entities. The integration of mixed reality (MR) and AR in XR technology along with the construction of this application economic system utilizing blockchain technology, secures the immersion of the user.

7.5.1.3 Interoperability

The metaverse incorporates various application scenarios and some of the functions can interact among themselves. Interoperability in Wang et al. (2022) can be demonstrated through the utilization of services like virtual sentiments and clothing that users have purchased in multiple metaverse settings.

7.5.1.4 Sociability

A new phase in the evolution of human social forms is the metaverse. Humans can seek higher-level demands beyond those of the physical realm, such as virtual work environments, entertainment, and other social settings. Thousands of fully autonomous people have contributed to building the metaverse and anyone can engage with anyone else at anytime, anywhere. For instance, if you signal or act, the other client will catch it right away and respond with feedback. The qualities of this interactive behavior are real time, precision, and diversity. As a consequence, sociality is a crucial component of user interaction in the metaverse.

7.5.1.5 Longevity

The metaverse is a digital universe that differs from the real world and has qualities of permanence. Avatars allow users to understand the importance of life's permanence in the metaverse. In other circumstances, the avatar's actions, possessions, and other information can be permanently imprinted and kept in the metaverse. To employ the AI algorithm to finish the "cloning" when necessary, the metaverse can fully capture the "gene" of the avatar through the retention and interpretation of user data. In contrast to the real world, many things in the metaverse vanish, along with the failure or death of an organization or a person.

7.5.2 Edge AI-based metaverse architecture

7.5.2.1 Edge cloud metaverse (ECM) architecture

The typical server-centric network design must transfer a lot of data between cloud and terminal devices in the metaverse. The network bandwidth is required to transmit data between the cloud server and terminal devices. However, it can change significantly over time due to the network's large-scale multiplexing characteristics. This causes a variety of issues, including network congestion and video frame dropouts, which harm the quality of user engagement. With the convergence of terminals, edge nodes, and cloud servers, the edge cloud architecture combines outstanding real-time performance, effective communication, and reliability. However, because users produce unbalanced data (i.e., statistical heterogeneity) and demand virtualized service types in various metaverse situations, this architecture comforts the problem of uneven distribution of both storage and computation resources at the edge nodes. The author in Said (2023) provides a metaverse paradigm that utilizes self-balancing federated learning (SBFL) to overcome the probabilistic and system diversity issues experienced by edge cloud architectures. In this design, the edge serves as a middleman to coordinate how users in various metaverse scenarios execute virtual services.

7.5.2.2 Mobile ECM architecture

As an emerging phenomenon for edge intelligence, mobile edge computing (MEC) deviates from conventional cloud computing that can be utilized on highly mobile devices like vehicle networks, mobile terminals, wearable technology, etc., that significantly decrease network latency by running as close to users as possible. For instance, the MEC into the metaverse successfully resolves the latency issue imposed on user mobility by merging several dynamic edge nodes to assist resources to the same user. However, when several edge nodes work together, it becomes necessary to send the user's data, which could result in privacy leakage and dire implications like crises of identity in the metaverse. To safeguard user data confidentiality in the metaverse, we suggest an FL-based MEC architecture in this meta-analysis. With this architecture, users can access the metaverse through intelligent internet of things (IoT) gadgets that produce data that is securely stored in the physical universe. Users' privacy is adequately protected because they only need to submit their local models to the FL layer and not divulge any other information. The development of metaverse features in physical and virtual world interactions is represented schematically in Figure 7.10.

7.5.2.3 Decentralized metaverse architecture

Through the metaverse, users can explore virtual worlds on their mobile smart devices and play in real time from any location. However, the growth of the metaverse poses scalability problems for the fundamental physical infrastructure (such as cloud servers and edge computing nodes). In other words, the infrastructure must handle more computational jobs as the metaverse's online user base grows. The server in Wienrich & Latoschik

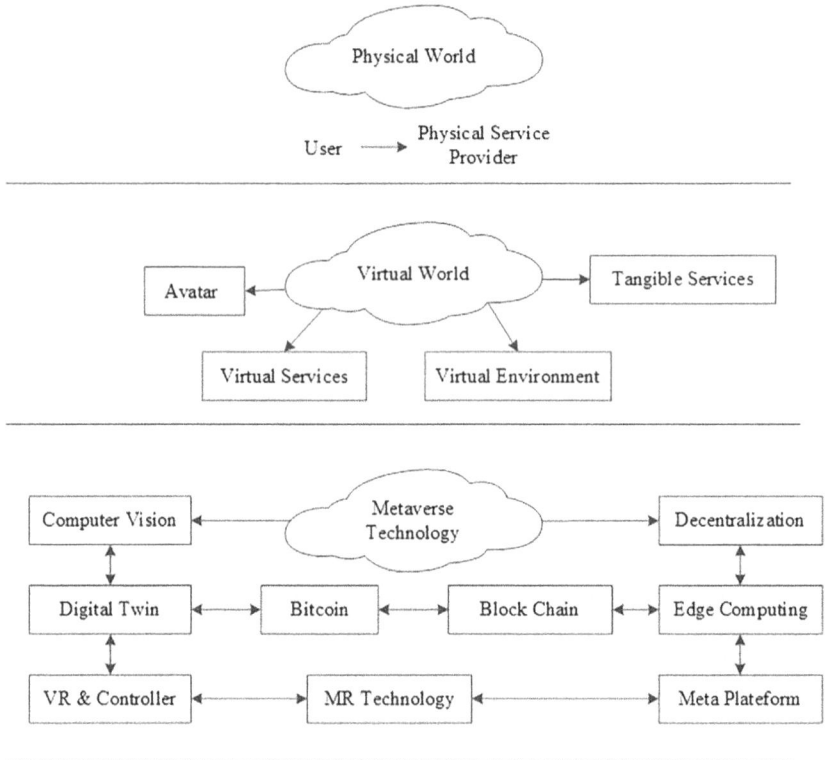

Figure 7.10 The development of metaverse features in physical and virtual world interaction.

(2021) must rapidly receive and reply to requests from additional terminals in a short time to maintain a high level of user experience. However, this is typically impracticable for the metaverse entities under conventional centralized architecture. However, the poor performance of the trained models limits the effectiveness of the aforementioned tactics. In particular, due to severe privacy restrictions, users in a blockchain-based decentralized metaverse network can only communicate with each other to train models which prevents them from efficiently utilizing as much user data as feasible. For this issue, we suggest a decentralized metaverse architecture built on FL and blockchain. Users with similar data distributions will deliver the task proposal disseminated by the blockchain once a user accesses a computing task through it. Then, these practitioners combine all of their local models and services onto the blockchain delivering them to the task applicant. A smart contract records the whole action on the blockchain for simply tracking and query by the concerned person and provider. Because of this, the local model is optimized under this architecture with the help of numerous user nodes, which successfully addresses the issue of inadequate model performance in the blockchain-based decentralized metaverse.

Consequently, the use of edge AI focused on 6G has the following three advantages:

- *Balance Data Storage:* In traditional cloud computing, the majority of the data produced by terminal devices must be transferred to the cloud for processing. However, the demand for cloud servers grows steadily heavier as more user devices are added. Edge intelligence uses load-balancing technologies to lessen redundant data storage by moving data produced by various terminal devices to edge storage nodes.
- *Efficient Data Transmission:* The authors in Yang et al. (2021) provide an architectural pattern of edge intelligence in the 6G that ensures faster transmission times than cloud computing channels, thereby lowering the demand for internet bandwidth, boosting the speed of information and communication transmission, and offering a platform for the quick response of virtual services.
- *High Reliability:* Future edge intelligence geared toward 6G will employ spatial multiplexing technology, where the likelihood of a communication disruption is less than one a million, considerably enhancing the user experience.

7.6 CHALLENGES AND APPLICATIONS OF EDGE AI IN 6G

7.6.1 Challenges of edge AI

7.6.1.1 Adversarial learning and adaptation

AI-enabled 6G will be confronted with a variety of cutting-edge threats that try to impair data analysis and decision-making. Therefore, it's vital to develop efficient self-adaptive strategies for recognizing and resisting these attacks (Hou et al., 2021). One possibility seems to be to incorporate the data impacted by the attacks as input and create a self-learning system that is resistant to various attack types.

7.6.1.2 Interpretable AI

Unfortunately, the majority of current AI solutions, particularly those based on deep learning, use a "black-box" approach without providing a clear justification for how the technique produced the specified result. One of the main obstacles to using AI in 6G systems is developing understandable AI with interpretable and predictable consequences.

7.6.1.3 Quality of experience

Instead of emphasizing quality of service, 6G is anticipated to concentrate more on optimizing and enhancing user quality of experience (QoE). The

QoE and user subjective experience are more directly connected. A generic and efficient model formulation that takes into account aforementioned factors and situations is yet to be developed to quantify the human users.

7.6.1.4 Interactive AI

The authors in Al-Ansi et al. (2021) provide the service functions in this circumstance that will depend on the hardware and software capabilities of each device as well as on the intelligence of those devices, which includes each user's reaction, capacity for learning, and speed of their interactions in the past and present.

7.6.1.5 Detecting and predicting human intention

The user traffic and demand that fluctuate in both time and space are unpredictable. For instance, an autonomous vehicle may regularly switch between manual mode and self-driving mode, resulting in significant changes in the data traffic linked to driving assistance.

7.6.1.6 Intelligent human-to-machine communications

The 6G era is expected to see the emergence of a universal and human-oriented networking framework in which various networking system components can sense, communicate, and socialize by the true intentions of human users, regardless of different interfaces, languages, and protocols.

7.6.2 Some more futuristic applications of edge AI in 6G

7.6.2.1 Industrial Internet of Things (IIoT)

The Internet of Things (IoT) refers to a network in which any object can be linked to the internet through internet-based protocols (Dubey et al., 2022). The primary objective of IIoT is to enhance the industry's performance. A distributed edge DL technique has been recently established to facilitate reduced latency semantic communication over IoT networks. This is accomplished by cooperatively optimizing the transmitters based on compressed deep neural networks. IIoT points to an industrial network that integrates industrial machinery and tools analyzes and interchanges the data generated, and optimizes production procedures. Industry 4.0 leverages digital twins as a significant technology for smart manufacturing by integrating physical entities with digital representations. Using a federated active transfer learning framework demonstrates the use of edge AI for IIoT applications.

7.6.2.2 Healthcare

To deliver a high standard of service in the healthcare industry, edge intelligence in 6G is anticipated to be extremely important. Edge intelligence can be used to fully utilize 6G's capability to provide 3D services, which will be useful for the healthcare industry. The works in Kamruzzaman, Alrashdi, & Alqazzaz (2022) elucidate the application of edge intelligence in healthcare. In Kamruzzaman, Alrashdi, & Alqazzaz (2022), the authors provided a thorough analysis of the brand-new possibilities, issues, and applications of edge intelligence in healthcare for smart cities. The edge node present in the edge technology continuously gathers, evaluates, and analyzes health data. Edge intelligence combined with decentralized and encrypted deep learning can significantly enhance the precision medicine and internet of medical things dependability, adaptability, preciseness, and confidentiality. To keep the medical data confidential, the researchers are currently focused on integrating blockchain with edge computing techniques. The prospective use of haptic communication in medical applications such as telerehabilitation, telediagnosis, and telesurgery eliminates needless hospital visits and saves time.

7.6.2.3 Autonomous driving vehicles

Autonomous vehicles are self-propelled vehicles that analyze the vast volume of data gathered by the onboard sensors in real time to constantly monitor their environs. Using multi-access edge computing with machine learning, 6G edge intelligence is expected to play a crucial role in enhancing the security and privacy of autonomous driving vehicles in the future. The two-tier wireless computing architecture benefits from edge learning for self-propelled vehicles addresses the benefits and difficulties of edge intelligence in connected autonomous vehicles. The authors in Mateescu (2023) discussed the potential application of edge intelligence in the domain of autonomous vehicle platooning where the autonomous vehicles move together in a coordinated or organized manner. Thus, it is identified that edge intelligence will play a crucial role in delivering ultra-low latency communication, analyzing real-time data, as well as enhancing security for autonomous vehicles.

7.6.2.4 Security and privacy

Enhancing security and developing a reliable and scalable edge AI system is of paramount importance for integrating connected intelligence in 6G. Nevertheless, edge networks are susceptible to privacy threats due to the enormous data exchange between the edge nodes and the scattered storage of information during the service operation. Numerous edge-learning models and architectures have been proposed, including FL, swarm learning, and split learning, to combat privacy leaks and adversarial attacks. The security and privacy challenges that go beyond 5G and developed a blockchain framework for integrating block chain technology with these systems, which guarantees both privacy and security. As edge applications beyond 5G networks require a varying level of privacy, the authors in Menon & Selvaprabhu (2022a) propose a multi-mode differential privacy scheme that adaptively applies data perturbation.

7.6.2.5 Education

Embedding edge intelligence in 6G allows ample and widespread support in the education sector. By making digital libraries, e-learning, remote learning, and teleconferences accessible with ease, intelligent edge computing in 5G transforms the way that education is delivered. The haptic technology, holographic media streaming, and in-place presence are all made possible by the 6G network's sophisticated edge computing. With holographic video conferencing and other technologies that are three-dimensional and life-sized, an in-place presence or streaming of holographic media can be achieved in education. Thus, it can be concluded that embedding edge intelligence in 6G is a cutting-edge technology that has the potential to reshape the education sector.

7.7 CONCLUSION

This chapter introduced the design and optimization of edge intelligence (EI) towards sixth-generation (6G) wireless communication. Edge artificial intelligence (AI) received considerable attention for enhancing reliability, low latency, low power, security, and trustworthiness, toward edge network users. The optical-free technology, quantum technology, native network slicing, integrated access backhauls networks, and holographic beam forming are some of the 6G technologies that enabled the development of wireless propagation solutions. Subsequently, this technology examines the impacts of various AI learning models, including federated, decentralized, split, distributed reinforcement, and trustworthy learning models. These edge network users achieved the maximum potential in data analytics by owing exploitation of edge AI architecture, including centralized, decentralized, hybrid, self-learning, and end-to-end edge AI architecture. The outcome of this architecture revealed the potential of improving the data normative aggregation and network performance, even among unknown services in the EI network. The deployment of edge cloud metaverse (ECM), mobile ECM, and decentralized metaverse architectures based on metaverse systems are demonstrated towards the ubiquitous of edge AI. Furthermore, a few key futuristic applications and challenges faced while utilizing the function of edge AI are discussed. Thus, it can be concluded that the EI is a potential contender for the futuristic 6G wireless communication systems. We hope this chapter will provide additional interest, guidance, and possible direction for researchers working in the emerging edge AI field.

We hope this chapter throws light on those researchers working in the emerging edge AI field.

REFERENCES

Al-Ansi, A., Al-Ansi, A. M., Muthanna, A., Elgendy, I. A., & Koucheryavy, A. (2021). Survey on Intelligence Edge Computing in 6G: Characteristics, Challenges, Potential Use Cases, and Market Drivers. *Future Internet*, 13(5), 118. 10.3390/fi13050118

Araujo Inastrilla, C. R. (2023). Internet Search Trends about the Metaverse. *Metaverse Basic and Applied Research*. 10.56294/mr202326

Chang, L., Zhang, Z., Li, P., Xi, S., Guo, W., Shen, Y., ... Wu, Y. (2022). 6G-Enabled Edge AI for Metaverse: Challenges, Methods, and Future Research Directions. *Journal of Communications and Information Networks*, 7(2), 107–121. 10.23919/jcin.2022.9815195

Chen, Y., Su, L., & Xu, J. (2017). Distributed Statistical Machine Learning in Adversarial Settings. *Proceedings of the ACM on Measurement and Analysis of Computing Systems*,1(2), 1–25.

Debauche, O., Elmoulat, M., Mahmoudi, S., Ahmed Mahmoudi, S., Guttadauria, A., Manneback, P., & Lebeau, F. (2021). Towards Landslides Early Warning System with Fog - Edge Computing and Artificial Intelligence**. *Journal of Ubiquitous Systems and Pervasive Networks*, 15(02), 11–17. 10.5383/juspn.15.02.002

Dong, P., Wu, Q., Zhang, X., & Ding, G. (2022). Edge Semantic Cognitive Intelligence for 6G Networks: Novel Theoretical Models, Enabling Framework, and Typical Applications. *China Communications*, 19(8), 1–14. 10.23919/jcc.2022.08.001

Duan, Q., Hu, S., Deng, R., & Lu, Z. (2022). Combined Federated and Split Learning in Edge Computing for Ubiquitous Intelligence in Internet of Things: State-of-the-Art and Future Directions. *Sensors*, 22(16), 5983. 10.3390/s22165983

Dubey, V., Mokashi, A., Pradhan, R., Gupta, P., & Walimbe, R. (2022). Metaverse and Banking Industry – 2023 The Year of Metaverse Adoption. *Technium: Romanian Journal of Applied Sciences and Technology*, 4(10), 62–73. 10.47577/technium.v4i10.7774

Foukalas, F., & Tziouvaras, A. (2021). Edge AI for Industrial IoT Applications. *IEEE Industrial Electronics Magazine*. 10.1109/mie.2020.3026837

Gong, Y., Yao, H., Wang, J., Li, M., & Guo, S. (2022). Edge Intelligence-driven Joint Offloading and Resource Allocation for Future 6G Industrial Internet of Things. *IEEE Transactions on Network Science and Engineering*. 10.1109/tnse.2022.3141728

Gupta, R., Reebadiya, D., & Tanwar, S. (2021). 6G-Enabled Edge Intelligence for Ultra -Reliable Low Latency Applications: Vision and Mission. *Computer Standards & Interfaces*, 77, 103521. 10.1016/j.csi.2021.103521

Hou, C., Hua, L., Lin, Y., Zhang, J., Liu, G., & Xiao, Y. (2021). Application and Exploration of Artificial Intelligence and Edge Computing in Long-Distance Education on Mobile Network. *Mobile Networks and Applications*. 10.1007/ s11036-021-01773-x

Jia, J. (2013). Joint Optimization for Congestion Avoidance in Cognitive Radio WMNs under SINR Model. *ETRI Journal*, *35*(3), 550–553. 10.4218/etrij.13. 0212.0297

Jiang, T., Shi, Y., Zhang, J., & Letaief, K. B. (2019). Joint Activity Detection and Channel Estimation for IoT Networks: Phase Transition and Computation-Estimation Tradeoff. *IEEE Internet of Things Journal*, *6*(4), 6212–6225. 10.1109/jiot.2018.2881486

Kalapothas, S., Flamis, G., & Kitsos, P. (2022). Efficient Edge-AI Application Deployment for FPGAs. *Information*, *13*(6), 279. 10.3390/info13060279

Kamruzzaman, M. M., Alrashdi, I., & Alqazzaz, A. (2022). New Opportunities, Challenges, and Applications of Edge-AI for Connected Healthcare in Internet of Medical Things for Smart Cities. *Journal of Healthcare Engineering*, *2022*, 1–14. 10.1155/2022/2950699

Lee, D., He, N., Kamalaruban, P., & Cevher, V. (2020). Optimization for Reinforcement Learning: From a Single Agent to Cooperative Agents. *IEEE Signal Processing Magazine*, *37*(3), 123–135. 10.1109/msp.2020.2976000

Legaard, C. M., Schranz, T., Schweiger, G., Drgoňa, J., Falay, B., Gomes, C., ... Larsen, P. G. (2022). Constructing Neural Network-Based Models for Simulating Dynamical Systems. *ACM Computing Surveys*. 10.1145/3567591

Li, E., Zeng, L., Zhou, Z., & Chen, X. (2019). Edge AI: On-Demand Accelerating Deep Neural Network Inference via Edge Computing. *IEEE Transactions on Wireless Communications*,*19*(1), 447–457.

Li, Y., Yu, Y., Susilo, W., Hong, Z., & Guizani, M. (2021). Security and Privacy for Edge Intelligence in 5G and Beyond Networks: Challenges and Solutions. *IEEE Wireless Communications*, *28*(2), 63–69. 10.1109/mwc.001.2000318

Lin, Y., & Shen, H. (2017). CloudFog: Leveraging Fog to Extend Cloud Gaming for Thin-Client MMOG with High Quality of Service. *IEEE Transactions on Parallel and Distributed Systems*, *28*(2), 431–445. 10.1109/tpds.2016.2563428

Liu, C. F., Bennis, M., Debbah, M., & Poor, H. V. (2019). Dynamic Task Offloading and Resource Allocation for Ultra-Reliable Low-Latency Edge Computing. *IEEE Transactions on Communications*, *67*(6), 4132–4150. 10.1109/tcomm.2019.2898573

Liu, Y., Yuan, X., Xiong, Z., Kang, J., Wang, X., & Niyato, D. (2020). Federated Learning for 6G Communications: Challenges, Methods, and Future Directions. *China Communications*, *17*(9), 105–118. 10.23919/jcc.2020.09.009

Mateescu, F. (2023). White Paper - Provisions on Artificial Intelligence. *Vector European*, 32–36. 10.52507/2345-1106.2023-1.06

Menon, V., & Selvaprabhu, P. (2022). Blind Interference Alignment: A Comprehensive Survey. *International Journal of Communication Systems*. 10.1002/dac.5116

Menon, V., & Selvaprabhu, P. (2022a). A Novel Tri-Staged RIA Scheme for Cooperative Cell Edge Users in a Multi-Cellular MIMO IMAC. *IEEE Access*, *10*, 117141–117156. 10.1109/access.2022.3219254

Rasheed, A., San, O., & Kvamsdal, T. (2020). Digital Twin: Values, Challenges and Enablers from a Modeling Perspective. *IEEE Access*, *8*, 21980–22012. 10.11 09/access.2020.2970143

Said, G. R. E. (2023). Metaverse-Based Learning Opportunities and Challenges: A Phenomenological Metaverse Human–Computer Interaction Study. *Electronics*, 12(6), 1379. 10.3390/electronics12061379

Selvaprabhu, P., Chinnadurai, S., Tamilarasan, I., Venkatesan, R., & Kumaravelu, V. B. (2022). Priority-Based Resource Allocation and Energy Harvesting for WBAN Smart Health. *Wireless Communications and Mobile Computing*, 2022, 1–11. 10.1155/2022/8294149

Tamilarasan, I., Selvaprabhu, P., & Venkatesan, R. (2023). Deployment of Hybrid FSO/RoF Links for 5G Heterogeneous Networks. *Journal of Optical Communications*, 0(0). 10.1515/joc-2022-0266

Tang, S., Zhou, W., Chen, L., Lai, L., Xia, J., & Fan, L. (2021). Battery-Constrained Federated Edge Learning in UAV-Enabled IoT for B5G/6G Networks. *Physical Communication*, 47, 101381. 10.1016/j.phycom.2021.101381

Wang, Y., Su, Z., Zhang, N., Xing, R., Liu, D., Luan, T. H., & Shen, X. (2022). A Survey on Metaverse: Fundamentals, Security, and Privacy. *IEEE Communications Surveys & Tutorials*, 25(1). 10.1109/comst.2022.3202047

Wang, Z., Xu, Y., Liu, J., Li, Z., Li, Z., Jia, H., & Wang, D. (2022). An Efficient Data Sharing Scheme for Privacy Protection Based on Blockchain and Edge Intelligence in 6G-VANET. *Wireless Communications and Mobile Computing*, 2022, 1–18. 10.1155/2022/5031112

Weng, Z., & Qin, Z. (2021). Semantic Communication Systems for Speech Transmission. *IEEE Journal on Selected Areas in Communications*, 39(8), 2434–2444. 10.1109/jsac.2021.3087240

Wienrich, C., & Latoschik, M. E. (2021). eXtended Artificial Intelligence: New Prospects of Human-AI Interaction Research. *Frontiers in Virtual Reality*, 2. 10.3389/frvir.2021.686783

Xiao, Y., Shi, G., Li, Y., Saad, W., & H. Vincent Poor. (2020). Toward Self-Learning Edge Intelligence in 6G. *IEEE Communications Magazine*, 58(12), 34–40. 10.1109/mcom.001.2000388

Yang, B., Cao, X., Xiong, K., Yuen, C., Guan, Y. L., Leng, S., … Han, Z. (2021). Edge Intelligence for Autonomous Driving in 6G Wireless System: Design Challenges and Solutions. *IEEE Wireless Communications*, 28(2), 40–47. 10.1109/mwc.001.2000292

Yang, K., Shi, Y., Zhou, Y., Yang, Z., Fu, L., & Chen, W. (2020). Federated Machine Learning for Intelligent IoT via Reconfigurable Intelligent Surface. *IEEE Network*, 34(5), 16–22. 10.1109/mnet.011.2000045

Yang, T., Qin, M., Cheng, N., Xu, W., & Zhao, L. (2022). Liquid Software-Based Edge Intelligence for Future 6G Networks. *IEEE Network*, 36(1), 69–75. 10.1109/mnet.011.2000654

Yuan, X., Angela Zhang, Y. J., Shi, Y., Yan, W., & Liu, H. (2021). Reconfigurable-Intelligent-Surface Empowered Wireless Communications: Challenges and Opportunities. *IEEE Wireless Communications*, 1–8. 10.1109/mwc.001.2000256

Zhou, Z., Chen, X., Li, E., Zeng, L., Luo, K., & Zhang, J. (2019). Edge Intelligence: Paving the Last Mile of Artificial Intelligence with Edge Computing. *Proceedings of the IEEE*, 107(8), 1738–1762. 10.1109/jproc.2019.2918951

Zhu, G., Lyu, Z., Jiao, X., Liu, P., Chen, M., Xu, J., … Zhang, P. (2023). Pushing AI to Wireless Network Edge: An Overview on Integrated Sensing, Communication, and Computation Towards 6G. *Science China Information Sciences*, 66(3). 10.1007/s11432-022-3652-2

Chapter 8

Artificial intelligence-based energy efficiency models in green communications towards 6G

Neelapala Anil Kumar[1] and Ravuri Daniel[2]
[1]Department of ECE, Alliance University, Bengaluru, Karnataka
[2]P V P Siddhartha Institute of Technology, Vijayawada, Andhra Pradesh

8.1 INTRODUCTION

The recent launch of 5G introduced high-throughput services to users in various nations, whereas 6G has been conceptualized by researchers globally (David et al., 2018). According to reports, 5G mobile devices and base stations (BSs) need a lot more energy than 4G devices. For instance, a 5G typical BS with diverse frequency bands consumes more than 11,000 W of power, compared to less than 7,000 W for a 4G BS. As the power utility raises the power amplification (PA) is enormous in multi-input multi-output (MIMO) antennas. The amount of 5G BSs required for identical service is now at a minimum for most instances of 4G, even though energy consumption per unit of information has fallen significantly. According to Information and Communication Technologies (ICT), services account for around 20% of all electricity usage and are expected to expand at an anticipated pace of 6% to 9% per year as per the statistics provided by (Andrae et al., 2015). Future THz-enabled BSs known as tiny base stations (TBS) can only really attain 100 m^2 (Lin et al., 2015). As the strategic sectors of such communication range can be lowered from 100 m of millimeter wave (mm-wave) spectral range to 10 m over THz spectrum, indicating that the expected number of BSs must increase considerably. Additionally, as illustrated in Figure 8.1, computational and content transfer services will gradually move from client PCs toward the cloud and peripheral systems, in addition to being significant aspects of ICT energy demand. Utilizing AI-based techniques is another crucial strategy for delivering excellent service, context-aware information transfers, and autonomous network management.

ICT infrastructure expansion, data explosion, and AI-based services will all result in increased energy usage (Wang et al., 2019) and (Mao et al., 2019). Both academia and industry have conducted a significant amount of research to lessen the rising energy burden associated with 6G. This research may be divided into two categories: deploying energy harvesting technologies and establishing energy-efficient network management algorithms.

DOI: 10.1201/9781003369028-8

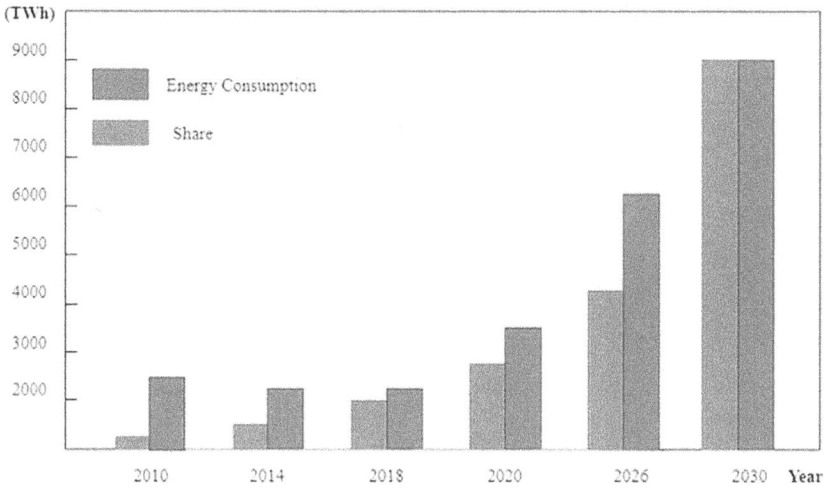

Figure 8.1 Energy consumption and its share.

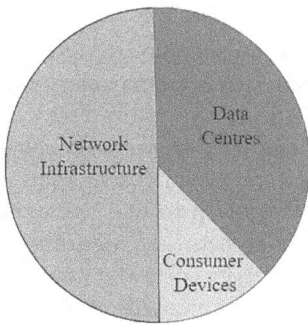

Figure 8.2 Energy consumption of different parts of ICT.

As shown in Figure 8.2, it is anticipated that energy-harvesting systems will be widely used to convert diverse sustainable energy sources into electricity for ICT devices, represented in Figure 8.3.

Additionally, sequential wireless communication and power transmission (SWIPT) and the utilization of interference signals are both made possible by Radio Frequency (RF) harvesting. Extensive execution numerical simulations have been provided to maximize system bit-per-Joule about energy-efficient network design (Feng et al., 2013). While to standardize the computational mathematics iteration process, advanced technologies, sometimes including Machine Learning (ML) approaches and conventional heuristic methods, have been implemented However, given that 6G network services need more than just quick throughput, tried-and-true methods for maximizing bit-per-Joule may not work. The creation of green communications must first satisfy a wide range of dynamic service requirements that are assessed using metrics other than bit-per-joule. To address this, Deep Learning (DL), a state-of-the-art

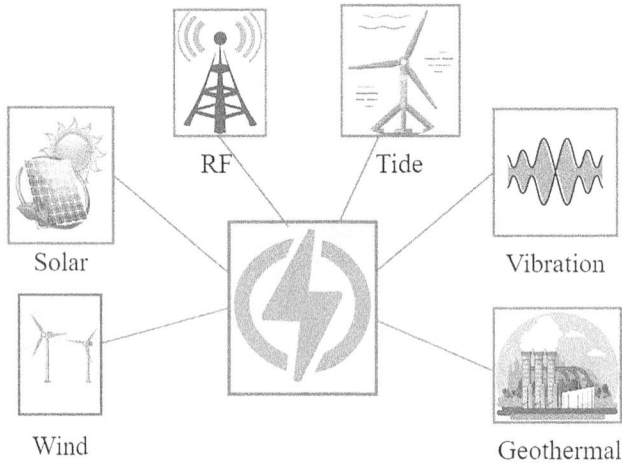

Figure 8.3 Different energy sources for ICT.

methodology, has been extensively studied in relation to the problem of resource network management (Zhou et al., 2018).

In 6G, numerous tactics were looked at because they may reroute the signal to the receivers for an enhanced signal-to-interference plus noise ratio (SINR), which has been evaluated as energy-efficient by many research studies (Liu et al., 2018). Unmanned aerial vehicles (UAVs) are now being considered to provide sporadic network access if energy-intensive continuous internet connectivity is not required. In the lead-up to 6G, we focus on AI models in this chapter, such as traditional heuristic algorithms, conventional ML, and cutting-edge DL, to reduce energy costs and improve energy efficiency. Moreover, unlike earlier studies that focused on specific networks by (Buzzi et al., 2016), we give succinct introductions to several 6G paradigms employed in existing works to clearly characterize works linked to green communication. To give readers some inspiration, we also outline future directions for 6G green communications research. Prospective 6G network scenarios, which are related to major 6G techniques and AI models, we predict issues with machine learning 6G green communications, such as massive computing complexity, security issues, practical deployment, and important 6G plans as well as the most popular AI methods for environmentally friendly communications and the related studies of greener CNC, MTC, and COC.

8.2 REVIEW ANALYSIS ISSUES TOWARDS GREEN 6G

This concept summarizes the currently available studies on green technologies with studies related to the issues of the 6G green environment.

8.2.1 Existing polls

Researchers have been interested in green communications subjects for more than a decade, according to relevant survey studies (Aziz et al., 2013). Concentration is on optical telecommunication services and providing comprehensive introductions to studies on energy consumption reduction for improving the energy efficiency of the physical layer in cellular networks, whereas (Erol-Kantarci et al., 2015) concentrate on the access layer. Moreover (Peng et al., 2015) offer tools for strategic consumption of power and architecture of energy efficiency. Fang et al. (2015) methodically present energy-efficient substitutes for crucial 5G technologies such as ultra-dense, enormous MIMO, and visible range communications (VLC). The wired and wireless elements of 3G, 4G, and 5G systems are examined in this study. Significant publications in recent history have examined energy efficiency challenges considering emerging technologies, which will be reviewed in this chapter. 6G monitoring equipment, including space air-ground integrated networks (SAGINs), machine learning, and non-orthogonal multiple access (NOMA), are anticipated to undergo multiple improvements. Energy harvesting has been covered in several assessments and does provide an intrinsic benefit for 6G. Although this tactic can dramatically lower the amount of energy that is consumed again from the electric grid, the dynamics of power harvesting raise security concerns. It has been discovered that extracting systems function is better when using AI-based strategies. Energy limitations have also been connected to several other problems, such as wideband, coding, and security (Chen et al., 2019). For non-convex or NP-hard challenges, collaborative optimizations of several metrics are often the goal of the research. Additionally, survey findings indicate that the often-used iteration-based strategies take a lot of time or sacrifice some performance. Future service requirements for 6G users will involve taking into consideration a variety of evaluation measures. Whereas AI will be the most practical and useful replacement due to the enhanced functionality and reduced processing and memory platform expected to become more extensively used to provide real-time computation and caching services, expertise connectivity might play a bigger role in 6G. Energy-saving techniques should consider simultaneous communications and computation/catching into account, even as computation and monitoring services require data communication. But prior studies have focused mostly on energy issues in communication systems. The most recent research into resource computation and prefetching services has not yet been thoroughly compiled and evaluated in relation to content-related communications.

8.2.2 6G research concerns towards 6G

From the information available, we found that there is a huge need for a thorough evaluation of green communications, considering the predicted

6G technology in earlier studies on energy-efficient communications. The study on cellular network communications (CNC), machine type communications (MTC), and computation-oriented communications (COC), as well as how it varies from earlier studies, are covered in this article. The energy efficiency of conventional study areas, including allocation of resources, servers' deployment, and routing, is also investigated by (He et al., 2019). To increase the study's interest, there is an in-depth survey on environmentally friendly communications. Figure 8.4 shows three scenarios that represent the main research topics as well as the obstacles to intelligent 6G.

1. **Communications via Cellular Network:** Communication over a cellular network green CNC research is largely concentrated on base station (BS) installation and work state scheduling because BSs utilize most of their energy in cellular networks. Additionally, user organization is crucial because it can be set up to switch off BSs when not in use. To increase energy efficiency, operational BSs should investigate power management and resource allocation, and energy conversion technology has been suggested to utilize less grid electricity. Contrarily, 4G and 5G will interact with 6G networks. This suggests that because of the hierarchy organization, the variety of BSs, and the heterogeneous spectral resources, it is more challenging. Additionally, regular BS reconfiguration will be necessary due to the whims of users, traffic, and gathered energy. The design of AI models will be challenging due to the unknowable channel state data (CSI), strong interference, and broad combinatorial optimization again for the power control problem (Yang et al., 2016).

2. **Communication between Machines (M2M):** Lowering energy consumption at the accessibility and core network should be taken into consideration because many MTC devices possess short battery lives and are difficult to recharge. The main topics of the study include access to the network, routing, and channel improvement. Energy harvesting is a crucial element of green MTCs; however, AI models' real-time effectiveness is hampered by unforeseen circumstances and unknown CSI. Additionally, we are driven to develop more effective AI models since the SAGIN architectural style will serve as the 6G paradigm for collectively delivering online access and information transmission with varied infrastructure.

3. **Computational-Oriented Communications:** In 6G, communication will be just as vital as computation and storage services. The study to save energy use mostly focuses on loading strategies and computing resource allocation because each server has a limited capacity. The location of edge servers and virtual machines (VMs), like BS deployment, is crucial to reduce energy use. Furthermore, cache

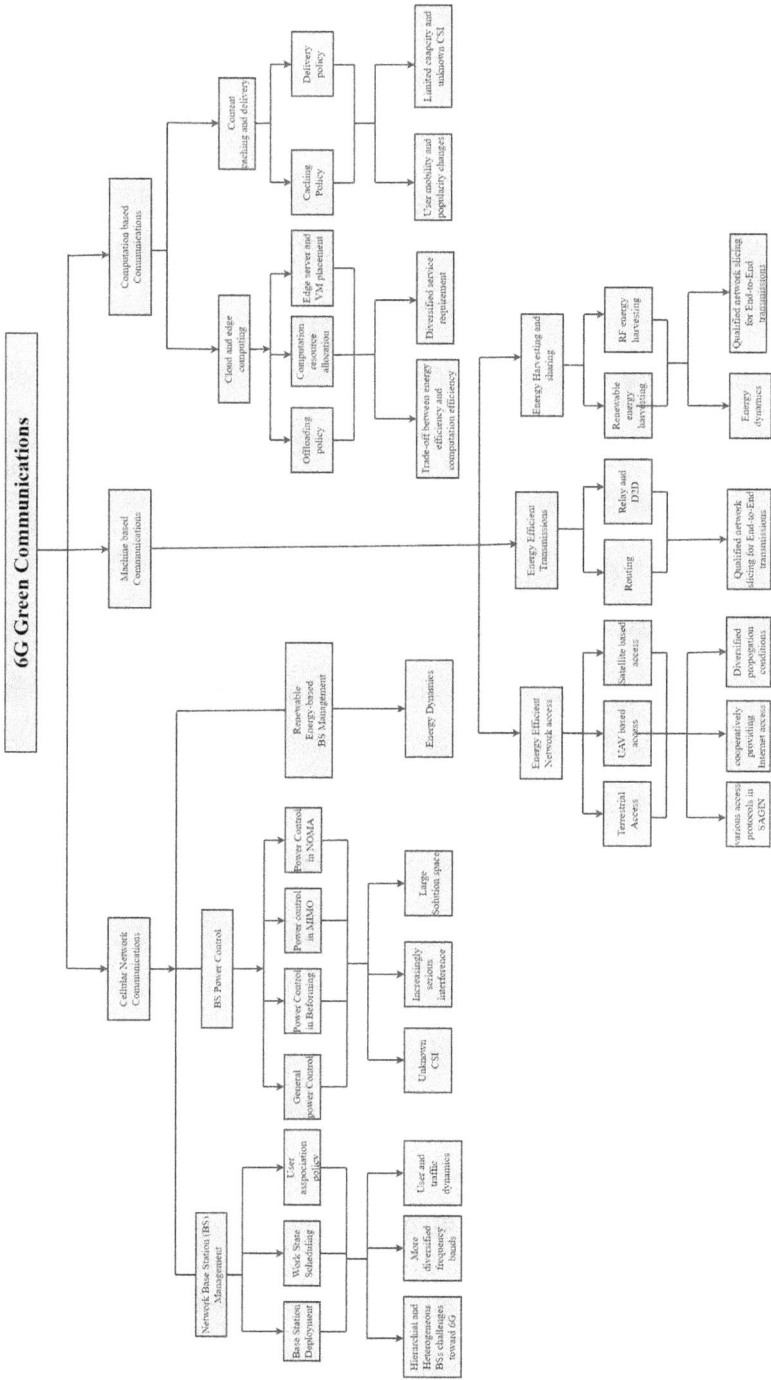

Figure 8.4 6G green communication solutions and challenges.

configurations for content distribution networks (CDNs) should indeed be optimized to limit content requests from faraway servers because the distribution process affects connectivity energy efficiency. However, AI models should consider several factors as networking has an impact on effective storage and computing facilities.

8.3 PARADIGMS OVERVIEW OF 6G AND ARTIFICIAL INTELLIGENCE METHODS FOR EFFECTIVE ENERGY COMMUNICATIONS

SAGINs, THz telecommunications, energy generation, and AI are a few of the main technologies usually referred to as 6G concepts. We will set up an additional perspective for these paradigms in this section. We will also go into more detail on the development of AI frameworks used during communications networks because this is a required subject on AI methods for green communications.

8.3.1 Several 6G paradigms

We will mostly discuss terahertz (THZ) connections, energy harvesting, and SAGINS since we will be emphasizing the building of AI models; we will just briefly explore AI-based communications as mentioned in the below subsections.

8.3.1.1 Terahertz communications

To achieve the optimum throughput of terabits of data per channel for 6G communications (Tbps), a wider bandwidth is needed. This is primarily seen in the sub-THz and THz frequencies. On the other hand, THz communications might be able to deliver the Tbps throughput and sub-1 ms end-to-end latency needed for the service. Contrarily, THz signals have a communication range of fewer than 10 meters, necessitating an increasing number of BSs to ensure flawless coverage, resulting in costly investment and energy costs. As a result, a greater emphasis should be placed on lowering energy overhead in the sustainable installation and operation of 6G BSs.

8.3.1.2 Space-air-ground integrated networks

As the human activity region expands farther and to more remote terrestrial sites, satellites and flying aircraft are being proposed to combine terrestrial communication networks to construct SAGINs. Furthermore, services of diverse quality can be offered to satisfy the various criteria due to the variety of communication facilities, channel capacity, and transmission lengths. Additionally, cooperative transmissions let users select the least energy-intensive routes to their destinations. On the other side, the variety of communication channels found in SAGINs increases the difficulty of optimizing network performance.

8.3.1.3 Energy harvesting

Energy recovery technology is expected to be widely used in 6G as wire-free charging methods advancements. Non-renewable sources are also included in the list of sources for harvesting, in addition to renewable energy sources like the sun, wind, and tide. Communication systems' reliance on electricity can be reduced by using renewable energy harvesting techniques. Non-renewable energy harvesting, in particular, RF-based harvesting can increase energy efficiency by recycling wasted signals. On the contrary hand, energy-gathering techniques make networks more unpredictable. To optimize system energy efficiency, robust methods for predicting energy available should be created.

8.3.1.4 AI-based communications

In addition to the classification of image tasks, natural-language processing, and game creation, AI techniques have been thoroughly studied to enhance network performance and green communication. Being a key application, the importance of AI as a paradigm for controlling 6G network automation has been confirmed. However, as network complexity rises and service standards tighten, existing AI systems face significant difficulties. The cooperation of several elements, including the design process, implementation, and resource utilization, among others, will determine the success of green networking management in the future. Many AI models have been created and put into use to realize intelligence in each component. But there are some innovative trends toward better communication management. We will introduce numerous current AI approaches, both traditional and contemporary, as well as potential AI models for the future.

8.3.2 Classical AI algorithms

The most utilized advanced technologies for green networking may be classified into three categories. Heuristic algorithms and traditional machine learning approaches are commonly used in traditional AI techniques. For some ML techniques, mainly heuristic algorithms are used by Rodrigues et al. (2020). For clear explanations, we solely treat non-data-based heuristic models; as a result, the former uses an online recursive-based search for the best responses but requires enormous amounts of data to gather experience to train definitive models. More details will be supplied in the next sections.

8.3.2.1 Heuristic algorithms

The non-deterministic polynomial-time hardness NP is the focus of heuristic approaches, which seeks to locate a workable solution quickly. Heuristic algorithms typically make use of many shortcuts and run more quickly than conventional greedy search methods. The trade-offs are decreased, seeking to maximize or close to global equilibrium. The heuristic methods that give rise to the shortcut techniques, as depicted in Figure 8.1, are particle swarm optimization (PSO), ant colony optimization (ACO) by (Bellavista et al., 2020), and genetic algorithm (GA), represented in Figure 8.5.

8.3.2.1.1 Optimization of particle swarms

In enhancing particle swarms, this optimization approach allows the named particles to traverse the search space using statistical interpretations of their position and velocity. Each particle's movement is influenced by its own best possible location along with the strongest spots in the search zone, leading to the identification of improved places. Repeating the process until a suitable answer is found. This technique has been employed to speed up the deployment of virtual machines (VMs) and edge servers (Venanzi et al., 2019). To improve energy effectiveness, the previously used method for continuous issues can also be used with a discontinuous process, as has been shown by Bellavista et al. (2020). In a high-dimensional space, this

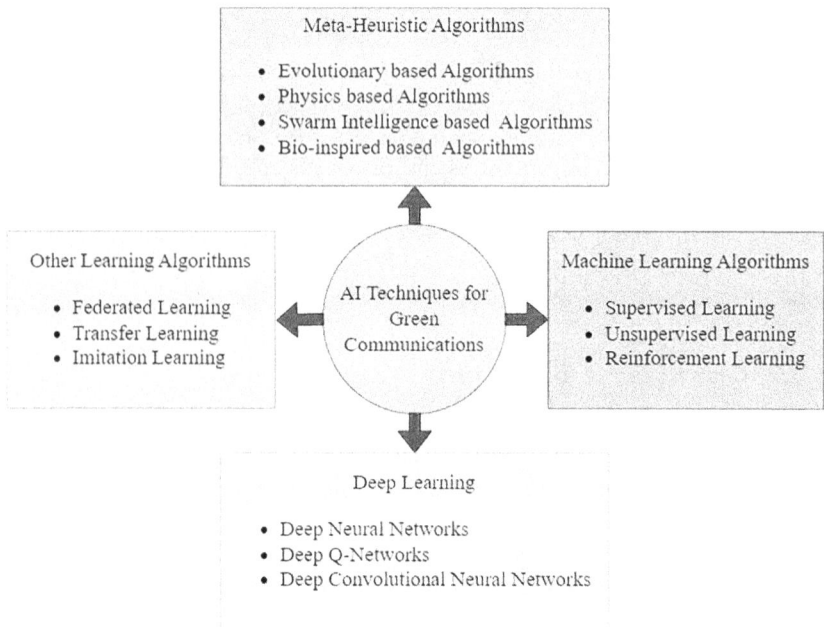

Figure 8.5 AI Techniques for green communications.

technique, however, has a limited rate of convergence and is prone to settling into the locally optimal.

8.3.2.1.2 Optimization of ant colonies

Ant colony optimization (ACO) was created to determine the best course by simulating the transformation, which was motivated by how ants search for food. Like PSO, ACO is built swarm-based, with a horde of "ants" (many computer-generated ants) scouring the search space for the best path. For other artificial ants to discover better locations during subsequent simulation sessions, note the location and characteristics of each artificial ant. To increase resource efficiency in a wide range of network tasks, involving routing (Liu et al., 2018), allocation of resources, and server deployment, this technique has been studied.

8.3.2.1.3 Genetic algorithm

Evolutionary concepts like polymorphism, crossover, and selection serve as the basis for genetic algorithms (GAs.) A set of potential answers is represented as chromosomes or phenotypes to improve the response in GA. A chromosome pair or gene mutations can cross across with a specific probability to create a new generation. Therefore, the mutation may appear in each succeeding generation, producing a chromosome or phenotype that is very distinct. To steer the process in the correct direction, fitness is constructed to evaluate people in each generation, and those with poor fitness values are removed. GA is easy to grow and consolidate, but it cannot ensure the optimum solution and is highly dependent on parameter selection. This technology has been used by scientists. This method has been used by researchers to optimize the deployment of edge servers to the moon and develop cellular networks.

8.3.2.2 Traditional machine learning

Several network performance optimization schemes have used ML algorithms as a data-driven strategy. Linear regression, SVM, and K-means grouping are the three popular machine-learning techniques utilized in green communications that are focused on in the coming sections.

8.3.2.2.1 Regression analysis

This method is frequently employed to investigate the relationship between several variables. A cost function is often created to assess the accuracy rate, and the labeled data set is typically utilized to map from the model parameters to the desired output outcomes. Multiple regressions can be separated into linear and nonlinear categories depending on whether the result is continuous or binary. both logistic regression and regression. For green communications to succeed, regression analysis is crucial. For instance, regression analysis can be used to predict future traffic patterns, as well as to identify resource allocation, transmission, and offloading strategies that are energy-efficient (Liu et al., 2018).

8.3.2.2.2 Support vector machine

Static vector machine (SVM) is employed in supervised learning to perform classification and regression analyses on data. An SVM model is given by a collection of orthogonal vectors. To divide the training data points, use a hyperplane or a group of hyperplanes. The hyperplane that is most like the training data is the optimal hyperplane in any class. High-dimensional issues with small data sets are best suited for SVM. SVM has been used in environmentally friendly communication to address problems including user association and computational burden.

8.3.2.2.3 Clustering with K-means

This approach divides many samples into clusters, assigning each observation to a cluster with the closest center. This methodology repeats the process of assigning nodes to clusters, as well as the cluster center, should be altered, as it is an unsupervised learning method. A cost function based on the separation among locations and the support vectors is constructed to assess the assignments. K-means clustering is useful for more and allocating them to suitable BSs to conserve energy. It may also be used to improve where cloudlets are located.

8.3.3 Deep learning

With developments in the ML training methodology, deep neural networks' predictive performance can be significantly increased. Furthermore, as depicted in Figure 8.1, deep learning (DL) uses the three common ML training techniques of supervised methods, unsupervised learning, and reinforcement learning (RL) methods. In this section, we will only give a few basic introductions to DL model construction, considering conventional DL features have already been introduced in several publications (Zhao et al., 2020). RL in DL, also referred to as deep reinforcement learning (DRL), is another topic we explore because it has more application than the other two training methods combined.

8.3.3.1 Development of deep learning models

Artificial neural networks (ANNs), also referred to as neural networks (NNs), and are the foundation for most contemporary machine learning models. ANN is made up of interconnected layers of "artificial neurons," which also are intended to resemble the neurons found in the human brain. Each artificial neuron could process data input, using nonlinear functions before sending the conclusions to neurons in the layer below via weighted edges. As a result, an ANN's final output is determined by both its nonlinear functions and edge weights in addition to its input signals. AI-predicted models on ANNs have expanded quickly in three areas over the past few decades. The larger number of layers results in deep structures, as compared to typical shallow ones, which is the most visible difference. Due to advancements in training algorithms and hardware advancements, current DL models can have highly complicated architectures while maintaining an extremely high accuracy rate. This enables them to be employed in very complicated situations and outperforms humans in some applications, such as board games. The linkages also become increasingly intricate. In certain modern ANNs, such as convolutional neural networks (CNNs), partial connections have been employed in addition to full interconnections between neurons in adjacent cells allowing for flexible evaluation of input with unevenly distributed features. Along with the creation of time-sequential variables. Third, researchers have created models, like the generative adversarial network (GAN) and actor-critical (AC) approach (Li et al., 2018), which allow numerous ANNs to work together to finish a job. While playing separate roles, the twin ANNs may use the same architecture or a different one. Additionally, a portion of the information can be used to retrain learning models.

8.3.3.2 Deep reinforcement learning

Dynamic learning techniques like reinforcement learning (RL) employ trial-and-error techniques to provide optimum results. The environment, a specified agent, the leading, the actions, and the reward are the primary elements of an RL model. In the investigated environment, the agent's decision is based on the situation at hand and is rewarded or punished for taking the right course of action. The agents either maintain previous experience or investigate a novel behavior with a high likelihood during the training stage to maximize the reward. The Q value, which reflects the anticipated accumulated reward for various actions taken at each stage in the conventional RL model, is frequently stored in a table. The aim of the training is to provide the Q-value table; however, it becomes unworkable as the examined issue becomes more complex unworkable to unviable. The fundamental idea of DRL is to be mapping from the condition to the appropriate action to remedy this issue. The ability to generalize the value of jurisdictions that an agent has never seen before is the advantage of this approach. Considering these advantages, DRL has received considerable attention to increasing energy efficiency by optimizing BS governance, resource allocation, power control, and compute offloading.

8.3.4 New training strategies

Changes in training methods have had a significant impact on correctness and computation performance. In addition to the creation of DL models, future networks will have more varied scenarios and dynamics; hence, a greater range of AI learning strategies will need to be investigated. This section concentrates on the collaborative learning and application learning AI training approaches, as seen in Figure 8.3.

8.3.4.1 Learning transfer

Transfer learning is a machine-learning technique that uses the created knowledge system to address many connected problems (Kwan et al., 2020). Transfer learning requires a perfectly all-right novel method depending on an existing information network or retraining a subset of it. This section, resulting in more complex and faster applications because of movement, of the network is constantly changing, making transfer learning appealing and analyzed for analogous situations. The present knowledge system's application space and the harmony between training and achievement in the goal scenario, on the other hand, are hot themes that need more investigation in the literature.

8.3.4.2 Collaborative learning

In this learning process, decentralized services or devices are utilized to train and evaluate AI models using local data, which is a form of decentralization. Because of this, edge devices can store training data locally and only upload the created characteristics to the centralized unit. The AI model parameters must be gathered and combined by the central controller. The peripherals can subsequently deploy AI models to generate forecasts or regularly update them. The growing rise in concern over personal privacy has increased interest in 6G federated learning. Additionally, federated learning's collaborative learning and running mode can make the best use of available edge computing resources, while using fewer central controllers. Furthermore, uploading variables is preferred to training data. Uploading parameters reduce connection overhead.

8.3.4.2.1 Summary

We can get to the following conclusions once the paradigm for 6G and AI techniques have been established:

The 6G standard will include significant advancements in THz communication, SAGIN, energy generation, and artificial intelligence.

Different algorithms and systems have been proposed, varying in performance and complexity.

To improve 6G performance, both conventional AI methods and state-of-the-art DL techniques can be used, depending on the complexity of the service requirement.

Most popular and effective AI models and methodologies have been analyzed in this area. Imitation training, random forest, and quantum machine learning are just a few AI techniques that have been applied in research. Additionally, some RL skills, including regret training and fictional learning, can be considered to increase performance. However, because of time restraints, it may be thought to increase effectiveness. We are unable to go into detail about each AI system and process, though, due to time limitations. There are engaging articles for readers. We go into further depth on how to use these strategies to create green communications in the section that follows.

8.4 OPEN RESEARCH PROBLEMS

Despite the many research projects on AI-based green data services, we need to place greater emphasis on translating our findings into useful applications for the 6G era. The use of AI techniques in contemporary networks is also complicated by issues with processing power, device compatibility, data protection, and other factors. We think the directions provided in the following paragraphs will be helpful to researchers.

8.4.1 Green BS management for 6 GNet

The BSs use most of the energy, which is anticipated to be numerous times more BSs within the 6G era than in the 5G period. Additionally, these BSs have different coverage zones and are built in a hierarchical fashion. Furthermore, hardware configurations and movement will put green management increasingly challenging since UAVs and high-altitude balloons (HABs) will be employed as BSs. The prospective AI-based study based on three probable 6G BS features is introduced in the following sections. Although different BSs, such as MBSs, SBSs, and TBSs, can provide terminal terminals in a multi-tier 6G HetNet, the overall user association strategy should be optimized to eliminate redundant BSs. Moreover, since BSs are frequently built with different frequency bands, channels and power allocation are essential for maximizing network energy efficiency. The BSs use most of the energy, while dynamic capability adds to the difficulty. The movement of edge devices and UAV or satellite-enabled BSs causes changeable traffic demands and dynamic resource situations. The use of intelligent systems can be very helpful in addressing these obstacles. To enable proactive network reconfigurations, AI models (Yang et al., 2019) can be employed to estimate traffic load, mobility patterns, and channel characteristics. Future BSs will also include computation/storage, energy, and other services in addition to communication. Because some BSs have constrained computer and storage resources, AI models can optimize content caching and computation offloading schemes. Examples of non-convex problems that need to be resolved to utilize RL or DRL methodologies include computational offloading and content caching. The RL or DRL can identify the globally optimal resolution while avoiding the time-consuming iteration phase during algorithm execution, in contrast to the conventional method, which divides the non-convex problems into two distinct ones and resolves them one at a time.

8.4.2 Low-energy space-air-ground integrated networks

In fact, SAGIN has been cited as a crucial 6G technology. SAGIN, especially for large MTCs, can offer seamless coverage and flexible information flow. Because many UAVs, HABs, and satellites are fueled by renewable energy, SAGIN is heavily dependent on energy-efficient network orchestration. On the other hand, the diverse hardware platforms, transmission conditions, and dynamic energy sources provide significant difficulties. AI can provide a variety of effective strategies for handling complexity and ambiguity. Utilizing the RL technique, for instance, to optimize resource allocation, such as transmission power and channels, might increase network energy efficiency. Using the RL approach, for instance, to optimize the resource allocation strategy for transmission power and channels. Energy-efficient packet transmissions are additionally made more challenging by CSI dynamics and network mobility. Because AI has shown it can quickly map the complex relationship between the existing network traces and potential terrestrial network transmission schemes. We believe the research can be extended to the SAGIN instance, even though improving SAGIN efficacy with AI has been investigated. Most current research focuses on single-line systems like low earth orbit (LEO) and UAVs. Network monitoring and supporting green communications should conduct a thorough investigation into each SAGIN component. UAV implementation and trajectory should indeed be adjusted with satellite beam control in mind to improve energy efficiency. Since AI has proven to be capable of solving complicated problems, using AI approaches to assess performance from the perspective of the entire SAGIN network will be a realistic strategy. How to describe the proper AI model components despite large computing expenses, and putting the AI model into action is another challenge.

8.4.3 AI-based energy-efficient transmissions

Packet transmission loses energy by requiring a lot of power from the transmitters, carriers, and recipients. Other energy-saving techniques have been presented in addition to power control and allocation of resources methods, such as routing policy formulation, relaying, scattering communications, and IRS-aided communications. There's no question that end devices will have access to a variety of communication channels for them to send packets successfully. Mobile users, for instance, can send emails over the cellular network using Wi-Fi that is based on IEEE 802.11 or D2D using a multi-hop approach. Network management and system energy consumption are significantly impacted by the use and cooperative management of numerous communication channels and resources in a have a significant impact on system energy use and network management inside multi-agent multi-task context. Only a small number of studies have looked at cases where more than one communication technique is used in AI-based research. In the long run, AI might be used to increase energy efficiency in situations where multiple communication modalities are present.

8.4.4 Artificial intelligence–enhanced energy harvesting and sharing

Energy harvesting is frequently seen as a crucial element of communications that are ecologically friendly. Green communications will be promoted using a variety of energy harvesting techniques, which may be divided into various groups depending on predictability and controllability, including the unregulated but foreseeable energy commission and the somewhat programmed energy groups, where its former incorporates alternative renewable power and RF energy and AI utility. For uncontrollable but foreseeable energy harvesting systems, a variety of AI models can be used to map the relationship among prospective reserves and related factors. The network can also be set up in advance using predictions. Utilizing AI models to directly link harvesting-related parameters to managed services policy is another approach. These tools help network administrators improve their knowledge of harvesting energy and boost resource utilization. AI can be utilized to improve the planning and regulation of BS power for the partially controlled RF harvesting approach. AI can be used to improve the trajectories of UAV-enabled BSs to reduce energy consumption and boost harvesting effectiveness. The focus of the current research is on increasing minimum harvests of energy because of chaotic transmissions and erratic power management (Kansal et al., 2007). AI has the potential to drastically reduce energy waste in RF harvesting, especially for signals from omni-directional antennas. Device-to-device energy sharing is another feature of RF harvesting technology that can be used to avoid network outages and cut down on power loss when a device's battery is fully charged and is no longer able to store incoming energy. The simultaneous wireless information (SWIPT) method has drawn a lot of interest, especially in MTC situations. AI can be applied to determine the proper ratio of RF harvesting to the transfer of information, even if it may cause some performance loss if energy is only extracted from a portion of the input variables (Perera et al., 2018). Currently, acoustic backscattering is indeed a promising technique, especially on low-power workstations, and AI may be utilized to improve information forwarding and resource harvesting.

8.4.5 AI-enabled network security

Online privacy is put at risk by adversaries and unauthorized users, who also creates transmission issues that decrease resource efficiency. AI can be used to defend against assaults by detecting network threats because it has been demonstrated to do so. To further combat network jammers, AI can be used to manage to transmit power and allot resources. Future AI-driven 6G networks may offer erroneous results due to a new form of network that exposes attackers' damaging data. Other probable outcomes include widespread end-terminal outages or incredibly low harvesting efficiency, in addition to decreased throughput or increased latency. The development of reliable AI models will be essential for ensuring green communications. The learning and execution phases of most AI systems, encompassing DL and ML, depend on data. Because the data may include sensitive personal or commercial information, data security should be considered when developing and implementing AI algorithms. More significantly, guidelines and standards for data gathering and utilization need to be created.

8.4.6 Design of a lightweight AI model and hardware

The power consumption of algorithms should be investigated to develop AI-based green communications. However, most current research neglects the energy needed to train and deploy AI models, concentrating instead on network performance enhancement over conventional methods. As a result, compared to conventional techniques, suggested AI technologies may be more complex and energy intensive. Therefore, for the development of intelligence green networking, lowering the training data amount required in addition to the algorithm complexities is essential. Communications may in some circumstances have lower average accuracy due to reduced complexity. The balance between energy consumption and network management is crucial for AI systems, even though decreased complexity may occasionally lead to decreased efficacy. Furthermore, device sensitivity affects how much energy AI algorithms use. Developing equipment for reduced AI algorithm computing optimization should receive more focus. There has only been a little amount of study done so far on how to run AI algorithms while using the least amount of energy. The results also motivate us to investigate ways to execute the recommended AI technologies in an energy-efficient manner.

8.5 SOLUTION FOR RESEARCH PROBLEMS

The performance of conventional communication systems is being seriously affected by the exponential rise of incremental data, fast throughput, and latency communication scenarios, which has posed substantial problems to the present communication methodologies. Recently, to enhance the efficiency of fifth-generation (5G) communications, non-orthogonal multiple access (NOMA), massive multiple-input multiple-output (MIMO), millimeter wave (mm-wave) technologies, and other appealing methods have emerged. Yet, according to IBM, the surging data will reach around trillion gigabits of 40 in 2020, a 44-fold surge since 2009, and it is anticipated that there will be 50 billion connected pieces of equipment by then. Importantly, new communication theories are necessary since existing communication techniques have basic limits in employing systems structure information and addressing massive amounts of data.

The creation of effective and dependable communication networks has been the focus of extensive research over the past few decades (Han et al., 2019). And others are focusing on the forthcoming generations. Let's use large MIMO as an illustration. A thorough investigation of the evolution of channel estimate methods using eigenspace, beam space, and angular velocity space has been studied. Unfortunately, complex eigen decomposition renders eigenspace-based channel estimate approaches challenging with huge antenna systems, which limits their application in upcoming gigantic MIMO systems. Continuous operation supported selection with the largest amplitude for transmission in the framework of beam-based channels estimation techniques, such as the priori-assisted (PA) channel tracking system (Gao et al., 2017), which resulted in enormous channel power leakage and catastrophic performance loss.

Angle space has drawn a lot of interest recently as just an angular domain-based channel to improve the effectiveness of channel estimation. Methods of estimation have been put forth. Even though the angle of departures and arrivals (AoAs/AoDs) follow a statistical distribution in a real-world setting, the previous studies assume that they reside at individual points in the angular domain. The assumption causes a power leakage issue in the angular space-based approaches as result, which lowers the effectiveness of the solutions. MIMO-NOMA has also been suggested as a viable technology that incorporates customer clusters, directional antennas, and power control with self-interference cancellation (SIC). However, there are still a lot of significant unresolved research questions in MIMO-NOMA, like the impact of SIC on the achievable data rate and trade-offs in decoding accuracy amongst NOMA-weak and NOMA-strong users. Although a variety of promising techniques enhance system performance, these frameworks frequently contain complex spatial structures and exceptionally intellectually stimulating NP issues, which restrict the practical use of these techniques.

The current progress of learning wireless physical layers is summarized in this article, along with various novels and effective learning-based communication systems. With a focus on 5G, three deep learning-based frameworks, NOMA, massive MIMO, as well as mm-wave hybrid precoding, were described as well as their performances were examined. It must be acknowledged, nonetheless, that several technical implementations still have many unanswered research problems, and it will take some time before deep learning ideas can be effectively applied to problems with the wireless physical layer.

Indeed, our research into this domain is still very young, and the intricate complexity makes advancement in this field challenging. Problems including model selection and data set acquisition, for instance, must be fixed. We need to develop the common sets of data that many individuals in the industry of wireless communication support, which would be good to construct explainable techniques for deep learning. We anticipate that our work will open a wealth of brand-new research questions for academics.

8.6 CONCLUSION

To improve quality, accelerate manufacturing, customize services, and other things, AI has sparked significant interest in almost every industry. Artificial intelligence's use in 6G is frequently viewed as a paradigm shift. AI-based sustainable communications will be a crucial path because of energy usage of infrastructure upgrades and end devices is expanding exponentially. Numerous factors and a wide range of effective solutions need to be examined and explored to reduce energy costs and increase energy efficiency. Traditional heuristic approaches and convex refinements may need numerous attempts or fail to achieve an appropriate level of energy efficiency, which are necessary to simplify the difficulties being addressed. AI techniques, on the other hand, AI approach have demonstrated great advantages and capacity to handle difficult problems. In this Chapter, we cited literature on network configurations and services for AI-related energy optimization. Utilizing regenerative braking technologies which use regenerative or ambient energy to cut down on the usage of fossil fuels is another method of communication that is environmentally friendly. To deal with the dynamism and uncertainties of the energy collection process, AI approaches can be used. This study also examines how AI may improve 6G element combinations such as large MIMO, NOMA, SAGIN, and THz communications in three well-known 6G circumstances: CNC, MTC, and COC. Our research, we hope, can help direct and inspire future studies on Intelligence 6G green communications. Additionally, we investigate the benefits and drawbacks of various AI strategies, including time-tested heuristic approaches and cutting-edge ML strategies. We show how they can cooperate methodically to lower energy use and boost energy effectiveness. Additionally, we stress the need of considering the energy consumption of AI models in addition to other pressing problems like privacy protection, computational complication, and equipment, and contributed to this growth that needs to be resolved by future researchers.

REFERENCES

Andrae, A., & Edler, T. (2015). On Global Electricity Usage of Communication Technology: Trends to 2030. *Challenges*, 6(1), 117–157. 10.3390/challe6010117.

Aziz, A. A., Sekercioglu, Y. A., Fitzpatrick, P., & Ivanovich, M. (2013). A Survey on Distributed Topology Control Techniques for Extending the Lifetime of Battery Powered Wireless Sensor Networks. *IEEE Communications Surveys & Tutorials*, 15(1), 121–144. 10.1109/surv.2012.031612.00124

Bellavista, P., Giannelli, C., & Montenero, D. D. P. (2020). A Reference Model and Prototype Implementation for SDN-Based Multi-Layer Routing in Fog Environments. *IEEE Transactions on Network and Service Management*, 17(3), 1460–1473. 10.1109/tnsm.2020.2995903

Buzzi, S., I, C.-L., Klein, T. E., Poor, H. V., Yang, C., & Zappone, A. (2016). A Survey of Energy-Efficient Techniques for 5G Networks and Challenges Ahead. *IEEE Journal on Selected Areas in Communications*, 34(4), 697–709. 10.1109/jsac.2016.2550338

Chen, Q., Wang, L., Chen, P., & Chen, G. (2019). Optimization of Component Elements in Integrated Coding Systems for Green Communications: A Survey. *IEEE Communications Surveys & Tutorials*, 21(3), 2977–2999. 10.1109/COMST.2019.2894154

David, K., & Berndt, H. (2018). 6G Vision and Requirements: Is There Any Need for Beyond 5G? *IEEE Vehicular Technology Magazine*, 13(3), 72–80. 10.1109/mvt.2018.2848498

Erol-Kantarci, M., & Mouftah, H. T. (2015). Energy-Efficient Information and Communication Infrastructures in the Smart Grid: A Survey on Interactions and Open Issues. *IEEE Communications Surveys & Tutorials*, 17(1), 179–197. 10.1109/comst.2014.2341600

Fang, C., Yu, F. R., Huang, T., Liu, J., & Liu, Y. (2015). A Survey of Green Information-Centric Networking: Research Issues and Challenges. *IEEE Communications Surveys & Tutorials*, 17(3), 1455–1472. 10.1109/comst.2015.2394307

Feng, D., Jiang, C., Lim, G., Cimini, L. J., Feng, G., & Li, G. Y. (2013). A Survey of Energy-Efficient Wireless Communications. *IEEE Communications Surveys & Tutorials*, 15(1), 167. https://www.academia.edu/2894477/A_Survey_of_Energy_Efficient_Wireless_Communications

Gao, X., Dai, L., Zhang, Y., Xie, T., Dai, X., & Wang, Z. (2017). Fast Channel Tracking for Terahertz Beam Space Massive MIMO Systems. *IEEE Transactions on Vehicular Technology*, 66(7), 5689–5696. 10.1109/TVT.2016.2614994

Han, S., Huang, Y., Meng, W., Li, C., Xu, N., & Chen, D. (2019). Optimal Power Allocation for SCMA Downlink Systems Based on Maximum Capacity. *IEEE Transactions on Communications*, 67(2), 1480–1489. 10.1109/TCOMM.2018.2877671

He, C., Hu, Y., Chen, Y., & Zeng, B. (2019). Joint Power Allocation and Channel Assignment for NOMA With Deep Reinforcement Learning. *IEEE Journal on Selected Areas in Communications*, 37(10), 2200–2210. 10.1109/JSAC.2019. 2933762

Kansal, A., Hsu, J., Zahedi, S., & Srivastava, M. B. (2007). Power Management in Energy Harvesting Sensor Networks. *ACM Transactions on Embedded Computing Systems*, 6(4), 32. 10.1145/1274858.1274870

Kwan, J. C., Chaulk, J. M., & Fapojuwo, A. O. (2020). A Coordinated Ambient/ Dedicated Radio Frequency Energy Harvesting Scheme Using Machine Learning. *IEEE Sensors Journal*, 20(22), 13808–13823. 10.1109/JSEN.202 0.3003931

Li, Y., & Wang, S. (2018). An Energy-Aware Edge Server Placement Algorithm in Mobile Edge Computing. 2018 IEEE International Conference on Edge Computing (EDGE). 10.1109/edge.2018.00016

Lin, Y.-D., Chu, E. T.-H., Lai, Y.-C., & Huang, T.-J. (2015). Time-and-Energy-Aware Computation Offloading in Handheld Devices to Coprocessors and Clouds. *IEEE Systems Journal*, 9(2), 393–405. 10.1109/jsyst.2013.2289556

Liu, C. H., Chen, Z., Tang, J., Xu, J., & Piao, C. (2018). Energy-Efficient UAV Control for Effective and Fair Communication Coverage: A Deep Reinforcement Learning Approach. *IEEE Journal on Selected Areas in Communications*, 36(9), 2059–2070. 10.1109/jsac.2018.2864373

Liu, J., Shi, Y., Fadlullah, Z. Md., & Kato, N. (2018). Space-Air-Ground Integrated Network: A Survey. *IEEE Communications Surveys & Tutorials*, 20(4), 2714–2741. 10.1109/comst.2018.2841996

Liu, X.-F., Zhan, Z.-H., Deng, J. D., Li, Y., Gu, T., & Zhang, J. (2018). An Energy Efficient Ant Colony System for Virtual Machine Placement in Cloud Computing. *IEEE Transactions on Evolutionary Computation*, 22(1), 113–128. 10.1109/tevc.2016.2623803

Mao, B., Kawamoto, Y., Liu, J., & Kato, N. (2019). Harvesting and Threat Aware Security Configuration Strategy for IEEE 802.15.4 Based IoT Networks. *IEEE Communications Letters*, 23(11), 2130–2134. 10.1109/lcomm.2019.2932988

Peng, M., Wang, C., Li, J., Xiang, H., & Lau, V. (2015). Recent Advances in Underlay Heterogeneous Networks: Interference Control, Resource Allocation, and Self-Organization. *IEEE Communications Surveys & Tutorials*, 17(2), 700–729. 10.1109/COMST.2015.2416772

Ponnimbaduge Perera, T. D., Jayakody, D. N. K., Sharma, S. K., Chatzinotas, S., & Li, J. (2018). Simultaneous Wireless Information and Power Transfer (SWIPT): Recent Advances and Future Challenges. *IEEE Communications Surveys & Tutorials*, 20(1), 264–302. 10.1109/comst.2017.2783901

Rodrigues, T. K., Suto, K., Nishiyama, H., Liu, J., & Kato, N. (2020). Machine Learning Meets Computation and Communication Control in Evolving Edge and Cloud: Challenges and Future Perspective. *IEEE Communications Surveys Tutorials*, 22(1), 38–67. 10.1109/COMST.2019.2943405

Venanzi, Riccardo, et al. (2019). Fog-Driven Context-Aware Architecture for Node Discovery and Energy Saving Strategy for Internet of Things Environments. *IEEE Access*, 7, 134173–134186. 10.1109/access.2019.2938888

Wang, Chien Ting, et al. (8 Dec. 2019). Cost Minimization in Placing Service Chains for Virtualized Network Functions. *International Journal of Communication Systems*, 33, e4222. 10.1002/dac.4222

Yang, Kai, et al. (1 Dec. 2016). Energy-Efficient Power Control for Device-To-Device Communications. *IEEE Journal on Selected Areas in Communications*, 34(12), 3208–3220. 10.1109/JSAC.2016.2624078

Yang, Qiang, et al. (2019). Federated Machine Learning: Concept and Applications. *ACM Transactions on Intelligent Systems and Technology (TIST)*, 10(2), 1–19.

Zhao, Yu, et al. (1 Dec. 2020). Lightweight Deep Learning Based Intelligent Edge Surveillance Techniques. *IEEE Transactions on Cognitive Communications and Networking*, 6(4), 1146–1154. 10.1109/TCCN.2020.2999479

Zhou, Yibo, et al. (Nov. 2018). A Deep-Learning-Based Radio Resource Assignment Technique for 5G Ultra-Dense Networks. *IEEE Network*, 32(6), 28–34. 10.1109/mnet.2018.1800085

Chapter 9

Centralized traffic engineering

M. W. Hussain

Department of Computer Science and Engineering, Alliance University,
Anekal, Karnataka, India

9.1 INTRODUCTION

There have been rapid developments in information and communication technologies (ICT) and thus the network speed is pivotal to sustain the newer applications. The current technology trends of 5G have become widespread with its improved frequency spectrum and integration of unlicensed and licensed bands (Chowdhury et al., 2020). Despite addressing the limitations, 5G still faces some intrinsic issues like integration of sensing, intelligence, communication, and control functionalities. Further, 5G is expected to reach its limits in 2030, so researchers must explore the avenues.

The constraints in 5G are well handled in 6G. 6G-based communication networks enable the newer services like human-machine interactions and terahertz communication (Tang et al., 2021; Ajibola et al., 2022). Further, there is a convergence of network densification, higher reliability and throughput, seamless connectivity and lower consumption of energy and it also includes newer technologies in addition existing functionalities in 5G. 6G communication networks include multitude of services like smart wearable devices, artificial intelligence, and Internet of Vehicles (IoV) and thus finds its suitability in smart cities (SCs). The most critical requirement for 6G wireless networks is the capability of handling massive volumes of data that emanate from the prolific sensors used in the multitude of devices and have very high-data-rate connectivity with stringent network requirements (Ajibola et al., 2022).

The former data emerging from the sensors requires massive computing resources. Cloud computing (CC) as a paradigm can provide higher computing requirements (like computing, storage, and network) on a pay-per-use basis. CC is suitable for applications viz. analytics, which requires batch processing and data analytics (Hussain & Roy, 2022; Hussain et al., 2021). Despite the higher computation requirements met by the CC, there is a higher latency to move the computation towards the cloud, which are prohibitive for meeting the quality of service (QoS).

The QoS of latency-sensitive applications which require immediate services and processing can be met by fog computing (FC), which provides

DOI: 10.1201/9781003369028-9

resources at the edge of the network in a time-centric approach. FC has limited resources compared to CC, but addresses the sensitive applications requirements. The proximity of the resources in FC for an application viz. healthcare, transportation system can meet the desired QoS and improves the much-needed performance of CC (Mukherjee et al., 2018). Further, mist computing (MC) finds its suitability as it is very close (1-hop) to the source and is leveraged for time-critical applications with pre-processing.

The stringent network requirements can be met easily while leveraging a software-defined network (SDN) (Hussain et al., 2020). SDN enables removing the control plane from the plane of the device, which provides dictating fine-grained network policy to be implemented in the network (Hussain et al., 2021). Thus, the centralized control plane can provide flexibility and removal of lock-ins imposed by the network vendor. However, as the network gets scaled up, a single controller is unable to meet the current and, thus, a multi-controller comes into action (Hussain et al., 2022).

The important issue faced by the networks today is how to tackle the challenges imposed by the large generation of data and to meet the desirable QoS of an application. This pertinent issue can be easily handled with the design of traffic engineering policy in the network for its optimization. Thus, traffic engineering policy implemented at the controller is the key to optimizing the performance for any application (Hussain & Roy, 2022).

This chapter discusses the traffic engineering policy to be implemented at the controller in SDN using the distributed computing paradigms viz. CC, FC, MC, etc., which handle massive computation requirements and networking requirements. The rest of this chapter is organized as follows: Section 9.2 discusses the traffic engineering in SDN. Section 9.3 discusses the multitude of computing paradigms. Section 9.4 discusses the intelligent transportation system utilizing CC, FC, and MC for a use case. Section 9.6 concludes this chapter and provides the efforts to be made in traffic engineering in the context of SD-IoV.

9.2 TRAFFIC ENGINEERING IN A SOFTWARE-DEFINED NETWORKING

This section introduces the basic notion of the software-defined networking (SDN) and NFV. It points out the related concepts of how the flows are set up, layered architecture of the SDN, and how SDN and NFV are jointly leveraged to address the network requirements, with an emphasis on the traffic engineering.

9.2.1 Flow setup in SDN

Traffic engineering implies the optimization of the network to improve network efficiency. Figure 9.1 illustrates the flows which are transmitted

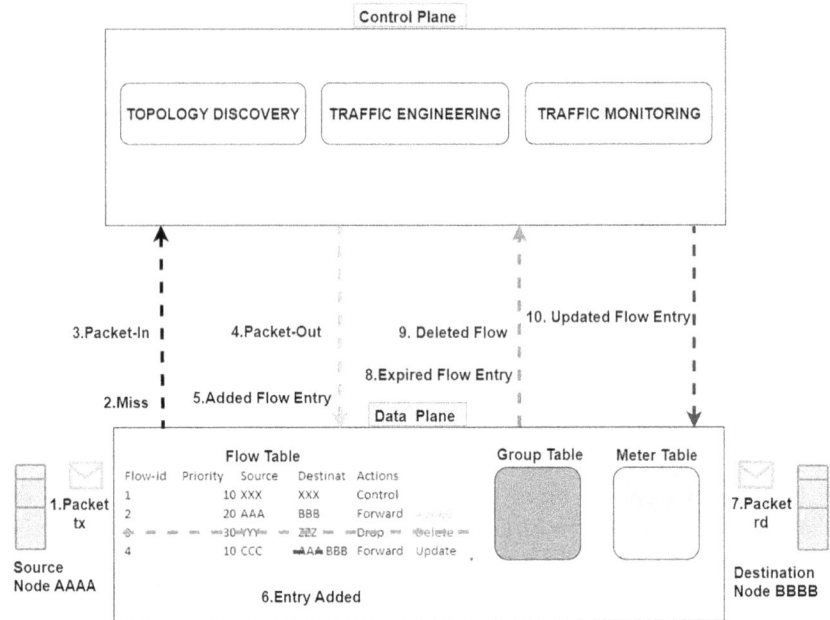

Figure 9.1 Setup of flows in software defined networking.

when the packets emanate from the network to the source and destination. Source node sends the packet to the nearby data plane. The data plane stores the packet information in the form of a flow table which captures several header information viz. source/destination address, priority, and the actions associated with it (Hussain et al., 2019). The actions along with each flow includes drop, control, and forward. SDN forwards the data using the flow table in contrast to destination-based routing in traditional networks and each flow is uniquely associated with the flow-id. In addition to the flow table being populated by the packets traversing the network, there is a meter table and group table. The meter table sets the minimum/maximum rate for data associated with each flow. The group table has a functionality to be used when the multicast/flood transmission is used by a source.

The data emanated from the source ingresses the data plane and the header information are leveraged to handle this packet. In this case, this is the first time the source sends a packet and thus to obtain the correct information how this flow is to be forwarded as it matches no flow inside the table, so a miss is generated and hence the data plane intimates the controller by generating a Packet_In (P_{in}) (Hussain et al., 2020). The controller maintains the overall information of the network in a centralized approach while leveraging OpenFlow protocol. Thus, the controller maintains the entire information of the network through reception of the information from the data planes operating the network. The reception of the P_{in} from the data plane will be replied to the data plane back via Packet_Out (P_{out}). Corresponding to P_{out}, there will be a flow entry installed in the data plane. The installation of flow entry will ensure the packet is forwarded. Each flow entry setup inside the flow time will have an expiry and when the time has elapsed, the successive entry will be expunged from the table. Data plane intimates the controller when it expunges the flow table after its expiry and the flow will be deleted from the table. Since the network is continuously changing so if there is any change in the network, the change is notified in the controller and thus the corresponding flow entry will be updated via the reception of flow update message from the controller. The centralization approach of SDN can be a perfect room to execute policies like load balancing, optimality routing, and slicing the network topology for the multiple tenants in a cloud platform. Thus, the continuous movement of data from the data plane and control plane ensures the controller knows the complete information about the network and thus a policy of traffic engineering can effectively be executed as per the network requirements.

9.2.2 SDN and network function virtualization (NFV)

As aforesaid, SDN logically separates the logical control and data plane. SDN has three layers and these three layers (data plane, control plane, and application plane), shown in Figure 9.2, communicate with each other via the application programming interfaces (APIs):

- *Application Plane:* This is the topmost layer of the SDN and has an emphasis on the plethora of network services operating the network. The application layer communicates with the control layer using northbound API. The application layer intimates the control plane to deploy the requisite network policies through the API.
- *Control Plane:* This is the middlemost layer of the SDN. It forms the necessary logic and receives the overall information from the lower layer using southbound API and is known as OpenFlow protocol. It is regarded as the brain of the network.
- *Data Plane:* This is the lowest layer of the SDN. It has the overall network physical infrastructure and will implement all the policies dictated from the control plane. This layer has a functionality of handling the overall data of the network by using the customized policies from the network as per the application requirements.

The separation of planes inculcates flexibility and avoids network ossification, which implies the prevention of network additions and upgrades. Thus, separation of planes prevents the higher operational and capital costs incurred in the network (Pradhan et al., 2022). NFV is a promising way to virtualize the functions of the network leveraging the proprietary network

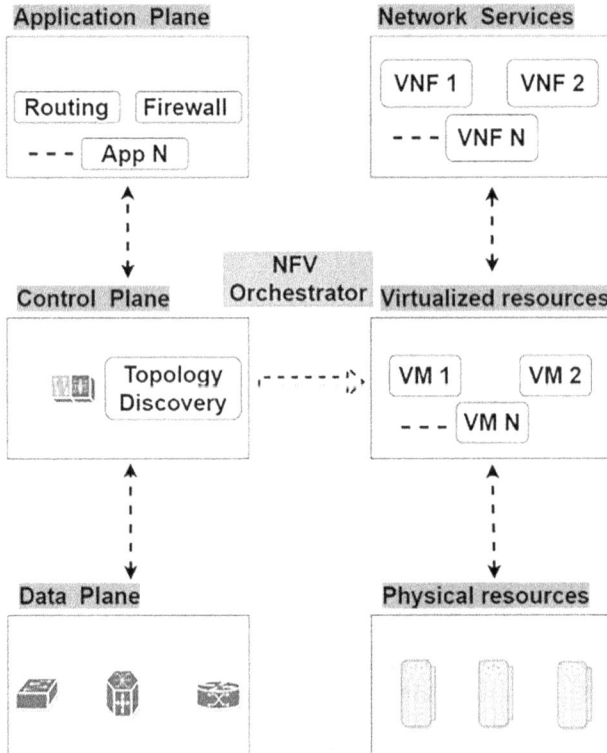

Figure 9.2 SDN and NFV.

hardware. So, NFV technology exploits virtualization technology to implement software functions in proprietary hardware and reduces operating and capital costs. NFV implemented in the network utilizes the commodity servers for implementation of virtual network functions (VNF) at lower cost. The physical infrastructure has a hypervisor running on the top of it to support the virtual machines. These virtual machines have running VNFs on the top of it. The logical component control module in Figure 9.2 is constituted by both the SDN controller and NFV. The provisioning of the VNFs is operated by the controller using standard APIs. The policy requirements from the application plane and the network topology gathered, then the control module executes the assigned network functions to the certain VMs and finally it is translated to the optimal routing. The function assignments are enforced by the NFV orchestration system, and the controller steers the traffic traveling through the required and appropriate sequence of VMs and forwarding devices by installing forwarding rules into them. Thus, with the integration of SDN and NFV, it facilitates reduction of the network costs, provisioning of networks, and addresses the traffic explosion, which improves the network.

9.3 COMPUTING PARADIGMS

This section provides us with the details pertaining to the myriad of paradigms that exist in the present world to address the application requirements. Also, merits and limitations of each paradigm has been discussed.

9.3.1 Cloud computing

Cloud computing (CC) is a promising paradigm from its inception because of the large advancements in ICT. The definition of CC as per the National Institute of Standards and Technology (NIST) is "CC presents an enabling platform that offers ubiquitous and on demand network access to a shared pool of computing resources such as storages, servers, networks, applications, and services". CC is suitable for high-performance computing like data analytics and batch processing (Ashu et al., 2020). CC presents a pay-per use model and can provision the resources efficiently and conveniently with ease of use (minimal interaction). CC has a centralized approach to computing and employs virtualization technology to create VMs and allocate to multiple users/tenants with the QoS guarantees through guaranteeing the service level agreements (SLA). Despite providing huge computing resources, cost-effectiveness through the pay-per-use model suffers due to the large latency for handling real-time data and time sensitive services of Internet of Everything (IoE) and 6G simultaneously (He et al., 2016; Chowdhury et al., 2020).

Due to generation of the huge amount of data by the proliferation of (IoT/IoE) sensors, these have stringent time requirements. As aforesaid, the centralized model of computing offered by the CC is insufficient with excessive bandwidth requirements, latency requirements, and resource-constrained devices (Ketu & Mishra, 2021). Thus, the huge demand of traffic emanated from sensors requires higher bandwidth requirements, stringent latency requirements of applications like vehicular communications, healthcare, and cloud resources are ineffective to be provided for resource-constrained devices like wearable devices.

9.3.2 Fog computing

The above limitations of the centralized paradigm are effectively handled with the use of fog computing (FC). As per the OpenFog Consortium, it is defined as "a system-level horizontal architecture that distributes resources and services of computing, storage, control and networking anywhere along the continuum from cloud to Things". It is a geographically distributed computing paradigm where a myriad of heterogeneous devices is connected at edge levels and offering flexible network, compute, and storage services. FC extends the CC near the edge of the network and is complementary to CC. FC is suitable for applications in which users require immediate services. FC is ideal for applications that require minimum data transmission rate and network latency with a throughput that is higher (Ketu & Mishra, 2021).

The issue with FC is that it takes several hops to reach the nearby fog server, which is itself prohibitive for data that requires pre-processing. With the pre-processing, only the relevant data will be moved to the next stage and hence can be a boon for resource-constrained devices.

9.3.3 Mist computing

Mist computing (MC) is much closer to the source of data and is ideally designed for time-centric applications. These time-centric applications will facilitate the improvement of higher throughput and low latency (Sharma & Park, 2021). With the help of MC, only data emanating from multiple devices will be pre-processed, which will enable only useful data to be moved further and thus help the processing time and bandwidth. MC offers multiple services for resource-constrained devices viz. power, communication, bandwidth, and memory. It also ensures the usage of resources in an optimal manner. The advantages of MC include quicker response time, light-weighted computing, and promising for local decisions (Ketu & Mishra, 2021).

9.4 INTELLIGENT TRANSPORTATION IN SMART CITIES

9.4.1 SD-IoV platform utilizing cloud and fog computing

In the current scenario of cities, a lot of problems exist owing to the higher population and higher urbanization. These pose a lot of issues, like social, technical, and economical, and are thus detrimental to the sustainability of modern cities. In consideration to this, the emergence of ICT and the myriad of sensors used for variety of cases has enabled the notion of a smart city. A smart city (SC) implies an urban area that aggregates data-utilizing sensors and electronic methods. Currently, the rapid growth has led to an increase in pollution, accidents, and traffic congestion, posing lots of difficulties to commuters and this has led to the new concept called intelligent transportation system (ITS), which is pertinent block of an SC (Sharma & Park, 2021; Ji et al., 2021).

Vehicles generate huge amounts of data and need real-time processing of data. Further, due to the continuous mobility of vehicles, there should be ubiquitous connectivity, higher data rates, and a focus on energy efficiency (Darabkh et al., 2022). These issues can be easily handled by leveraging 6G, which also provides terahertz communication. The IoV are a pertinent application of Internet of Things (IoT) technology in the field of the ITS. IoV extend the paradigm of vehicular adhoc network (VANET) but dissects from the VANET owing to the centralized management making it promising for ITS. IoV are enabled by a multitude of wireless-based sensors for providing a safer, effective, and better quality of experience (QoE) to the customers. Furthermore, the vehicles also include GPS-based sensors for providing navigation, sensors for monitoring the pollution because of the combustion, etc. to make the vehicles more desirable for smart cities. Vendors for the automobile industry provide the vehicles with on-board units (OBUs) to support the plethora of vehicular applications. These applications include several infotainment applications like online gaming, social media-based applications, etc. The applications of vehicular networks require higher computation power, storage, and are delay sensitive. The IoV not only carries on wireless communication between vehicles and vehicles, roads, pedestrians, and the internet, but also realizes intelligent traffic management, intelligent dynamic information service, and intelligent vehicle control network integration. Various services are expected to be provided in the IoV, such as collision warning, traffic congestion detection, route planning, infotainment, etc., which makes human traffic travel more convenient. Despite the OBU present in the vehicle, the higher frequency of the data emanated from the sensors are incapacitated to handle such computation. Thus, to provide substantial resources to the vehicle for facilitating robust vehicular networks, a new paradigm must emerge.

As stated above, IoV generate a large amount of data and being latency sensitive data needs stringent QoS requirements. The huge computing resources required for the IoV are met by the cloud computing platform. Cloud computing (CC) is a promising platform to tackle the huge computing requirements. The cloud plays a pivotal role which follows a pay-per use model. Thus, the cloud provides an enabling platform to provide higher computation and storage needs but comes at the cost of higher propagation delay, which should be avoidable for delay-sensitive tasks in IoV. To avoid the excessive cost of the centralized cloud platform, computation resources should be provided near the proximity of the vehicles and, thus, FC comes into action. FC acts complementary to CC, where the resources are provided nearby to the IoV. FC can address the requirements of the IoV by providing substantial computational resources closer, while maintaining the necessary deadlines.

However, the transportation of data from the IoV to either FC or CC also has stringent network requirements. Traditional networks employing both control and data plane simultaneously at a single device are incapable of addressing the network requirements due to vendor lock-ins thus limiting innovation in the network. SDN is a new paradigm in the domain of networking that has gained prominence in recent years with the mass scale deployment of complex systems owing to its unique and flexibility features (Hussain & Roy, 2021). SDN dissects the control plane from the data plane and thus implements a flow model for every data transmitted. This flow-based model is complementary to destination-based routing in traditional networks and allows a customized policy to be enforced as per the

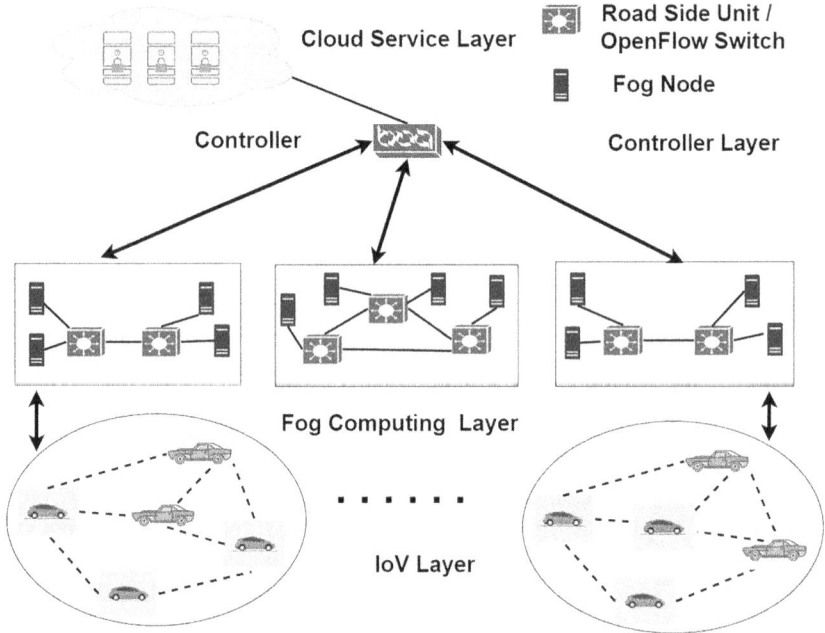

Figure 9.3 SD-IoV in ITS utilizing cloud and fog computing.

networking requirements. Thus, SDN is regarded as a boon for the next generation applications due to the dissection of control plane from the data plane. Further, the continuous mobility of vehicles adds to the problem of ascertaining topology and can easily be guessed with the help of SDN. Figure 9.3 discusses the SD-IoV, as a use-case of transportation utilizing both CC and FC. OpenFlow switches (OFSs) also act as a road side unit (RSU), as described in Figure 9.3. Therefore, the higher requirements of both the computation needs and network requirements, while taking note of the mobility of vehicles in IoV in the present day, can easily be met by the distributed FC platform and the centralized feature of SDN.

9.4.2 SD-IoV platform utilizing mist, cloud, and fog computing

SD-IoV face a lot of issues viz, identification of real-time events plying on the roads within the deadlines. To utilize the above platform for considering this scenario, it might be prohibitive because fog nodes are a few hops away from the generation of data. Since the data upload from the source of data might take substantial amount of time, so a novel mist node is incorporated near the base stations (close or one hop away from the vehicles). Leveraging of mist nodes in the SD-IoV facilitates the preprocessing of data prior to data being sent to the fog nodes. With the

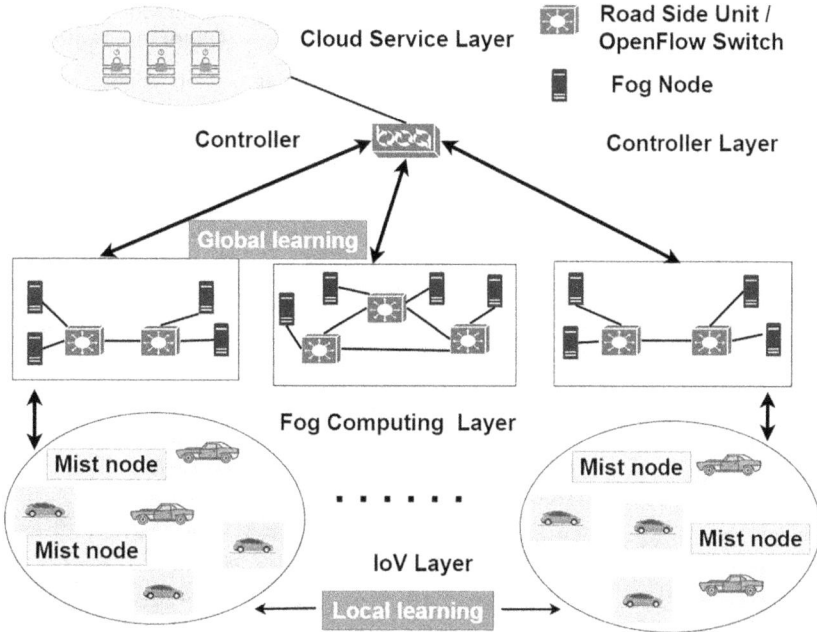

Figure 9.4 SD-IoV in ITS learning model utilizing cloud, fog, and mist computing.

pre-processing of data in the mist nodes, the network load should be minimized and besides, with the information emanated from the source-mist nodes, a proper machine learning model can be trained to predict the events in a timely manner. Although mist nodes have a much less flux of data, the same model can also be shared with the fog nodes that are distributed across the network and will be trained with the global shared model in the fog nodes. Figure 9.4 describes the learning model in SD-IoV leveraging CC, FC, mist nodes, and SDN while using 6G networks for predicting the events on the roads. Thus, the crucial time events can be predicted within deadlines, while the prediction accuracy can easily be improved from training the local mist model with the fog model utilizing the benefits of a centralized SDN controller.

9.4.3 Topology based slicing in SD-IoV platform

As aforesaid, NFV plays a pertinent role in virtualizing the network functions and SDN controller plays an important part in ensuring TE in the network. Controller is the focal point of the network and provides any abstraction in the network. The topology-based slicing implies creation of different parts of the network to myriad of operators. With this slicing, each network operator utilizing the network will have a slice of it and thus abstracted from the other users by creating a virtual environment of the

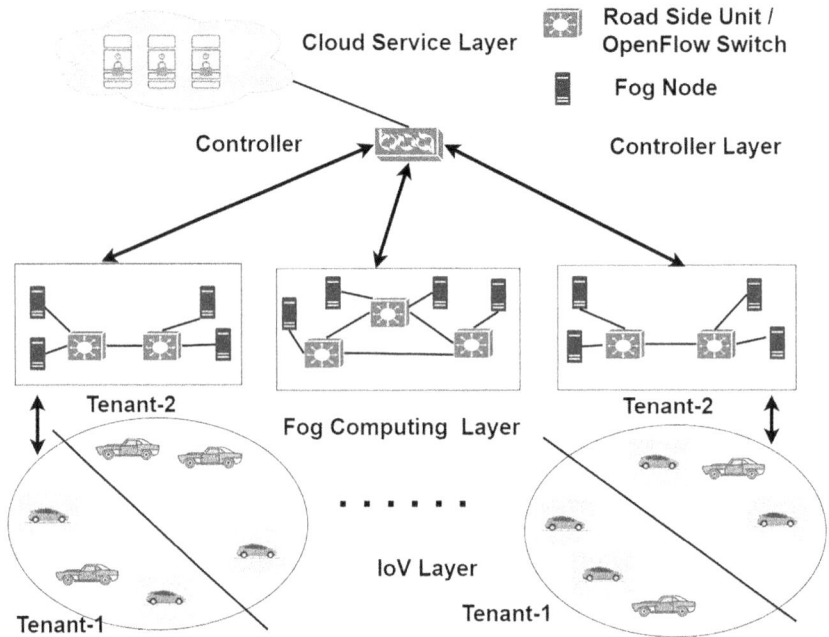

Figure 9.5 Topology-based slicing in SD-IoV in ITS.

entire network. The centralized controller also acts as the fog orchestrator, so the controller acts as a twofer for both the slicing of the network and allocation of the resources for the vehicles at fog nodes (Phan et al., 2021). The controller also performs another functionality of load balancing that distributes the resources among the fog servers uniformly.

The traffic handling during slicing is handled by creation of flow space in each of the data planes (Khan et al., 2017). The network load is substantially reduced by focusing on specific data planes. The slicing of the network is itself dependent on the ports and location of the data plane present in the network. Figure 9.5 describes the topology-based slicing among multiple tenants in SD-IoV. Thus, with topology-based slicing in the centralized controller SD-IoV, the controller can ensure multiple tenants to operate the same network and isolate the traffic with each other to avoid any outages/mix of the traffic and this enables it to provide a dedicated slice to the traffic with high priority.

9.5 OPEN ISSUES

Transportation has always been the main concern of mankind due to the several issues; however, many of them have been addressed by SD-IoV. Despite overcoming intrinsic issues in transportation, still some issues persist like QoS maintenance, network outages due to the continuous mobility of vehicles, and battery charging. 6G can well handle many issues because its foundation relies on the usage of terahertz communication, the integration of the energy transfer, and wireless information slicing network in a dynamic manner, etc., thus maintaining the requisite QoS.

The data emanated from the vehicles is huge and with the mobility of vehicles, large amounts of data are to be transferred, thus overburdening the network. This huge mobility and large transfers can be well accommodated by the centralized controller. Since the controller populates the flow table of the OpenFlow switch continuously and due to the large number of flows, it might increase the interaction time; hence, protocol-independent packet processors (P4) can come in handy to minimize the interaction time and improve the performance of the SD-IoV. Despite the huge data, a single controller might not handle the huge resources, as it can be a single point of failure; thus, the multi-controller comes into the picture. However, with the use of multi-controllers, both the resource allocation/sharing becomes the main problem due to the architecture design (flat/hierarchical) of the multi-controller and the non-standard communication of eastbound/westbound API with the other controllers. Further, the placement of the controller can also play a pivotal role in reducing the latency.

The processing of the data emanated from the vehicles (time-sensitive services) has to be processed within deadlines. In this regard, MC can play a crucial role by pre-processing the data rather than directly transferring data to the fog node. The use of crucial technologies like machine learning and artificial intelligence (AI) can facilitate the improvement of the SD-IoV for improving accuracy and thus deciding what actions can be taken when data is received from the vehicle sensors.

Further, the data emanated from the sensors is huge and has stringent QoS requirements, so the network devices need to be flexible and open. As aforesaid, SDN has the power to ensure dynamic routing (can be designed network-specific protocols and operate with traditional network protocols simultaneously) and has a well-defined API. The use of well-defined API can ensure device and network programmability, which is a core feature of any 6G network. Thus, the above-mentioned open issues can improve several issues plaguing the transportation sector utilizing SD-IoV.

9.6 CONCLUSION

This chapter discusses the core issue of centralized traffic engineering in SDN utilizing 6G networks. In this context, a use case of transportation utilizing SD-IoV in an SC is discussed, which is central to improve the lives of the people. This use case is handled using distributed computing multiple paradigms like CC, FC, and MC, which provide several benefits. The issue of traffic engineering is to be managed by the controller through load balancing, NFV, network slicing, and fog orchestrator to improve the network utilization. The open issues are also discussed to further the avenues of this research, like the usage of a multi-controller when the OFSs are scaled up, programmable data planes to improve the controller scalability, controller placement problem and implementation of ML models, and AI can further improve the SD-IoV.

REFERENCES

Ajibola, O. O., El-Gorashi, T. E. H., & Elmirghani, J. M. H. (2022). Disaggregation for Energy Efficient Fog in Future 6G Networks. *IEEE Transactions on Green Communications and Networking*, 6(3), 1697–1722. 10.1109/tgcn.2022. 3160397

Ashu, A., Hussain, M. W., Sinha Roy, D., & Reddy, H. K. (2020). Intelligent Data Compression Policy for Hadoop Performance Optimization. In *Proceedings of the 11th International Conference on Soft Computing and Pattern Recognition (SoCPaR 2019)*, 11, pp. 80–89. Springer International Publishing.

Chowdhury, M. Z., Shahjalal, Md., Ahmed, S., & Jang, Y. M. (2020). 6G Wireless Communication Systems: Applications, Requirements, Technologies, Challenges, and Research Directions. *IEEE Open Journal of the Communications Society*, 1, 957–975. 10.1109/ojcoms.2020.3010270

Darabkh, K. A., Alkhader, B. Z., Khalifeh, A. F., Jubair, F., & Abdel-Majeed, M. (2022). ICDRP-F-SDVN: An Innovative Cluster-Based Dual-Phase Routing Protocol Using Fog Computing and Software-Defined Vehicular Network. *Vehicular Communications*, 34, 100453. 100453.10.1016/j.vehcom.2021. 100453

He, X., Ren, Z., Shi, C., & Fang, J. (2016). A Novel Load Balancing Strategy of Software-Defined Cloud/Fog Networking in the Internet of Vehicles. *China Communications*, 13(Supplement2), 140–149. 10.1109/cc.2016.7833468

Hussain, M. W., & Roy, D. S. (2022). Intelligent Node Placement for Improving Traffic Engineering in Hybrid SDN. In S. Dhar, S. C. Mukhopadhyay, S. N. Sur, & C.-M. Liu (Eds.), *Lecture Notes in Electrical Engineering*. Springer Singapore. 10.1007/978-981-16-2911-2

Hussain, M. W., & Sinha Roy, D. (2021). Enabling Indirect Link Discovery Between SDN Switches. *Proceedings of the International Conference on Computing and Communication Systems*, 471–481. 10.1007/978-981-33-4084-8_45

Hussain, M. W., Reddy, K. H. K., & Roy, D. S. (2019). Resource Aware Execution of Speculated Tasks in Hadoop with SDN. *International Journal of Advanced Science and Technology*, 28(13), 72–84. http://sersc.org/journals/index.php/ IJAST/article/view/1282

Hussain, M. W., Pradhan, B., Gao, X. Z., Reddy, K. H. K., & Roy, D. S. (2020). Clonal Selection Algorithm for Energy Minimization in Software Defined Networks. *Applied Soft Computing*, 96, 106617. 10.1016/j.asoc.2020.106617

Hussain, M. W., Reddy, K. H. K., Rodrigues, J. J. P. C., & Roy, D. S. (2021). An Indirect Controller-Legacy Switch Forwarding Scheme for Link Discovery in Hybrid SDN. *IEEE Systems Journal*, 15(2), 3142–3149. 10.1109/jsyst.2020. 3011902

Hussain, M. W., Khan, M. S., Reddy, K. H. K., & Roy, D. S. (2022). Extended Indirect Controller-Legacy Switch Forwarding for Link Discovery in Hybrid Multi-Controller SDN. *Computer Communications*, *189*, 148–157. 10.1016/j.comcom.2022.03.017

Ji, T., Chen, J.-H., Wei, H.-H., & Su, Y.-C. (2021). Towards People-Centric Smart City Development: Investigating the Citizens' Preferences and Perceptions About Smart-City Services in Taiwan. *Sustainable Cities and Society*, *67*, 102691. 10.1016/j.scs.2020.102691

Ketu, S., & Mishra, P. K. (2021). Cloud, Fog and Mist Computing in IoT: An Indication of Emerging Opportunities. *IETE Technical Review*, *39*, 713–724. 10.1080/02564602.2021.1898482

Khan, S., Gani, A., Abdul Wahab, A. W., Guizani, M., & Khan, M. K. (2017). Topology Discovery in Software Defined Networks: Threats, Taxonomy, and State-of-the-Art. *IEEE Communications Surveys & Tutorials*, *19*(1), 303–324. 10.1109/COMST.2016.2597193

Mukherjee, M., Shu, L., & Wang, D. (2018). Survey of Fog Computing: Fundamental, Network Applications, and Research Challenges. *IEEE Communications Surveys & Tutorials*, *20*(3), 1826–1857. 10.1109/comst.2018.2814571

Phan, L.-A., Nguyen, D.-T., Lee, M., Park, D.-H., & Kim, T. (2021). Dynamic Fog-to-Fog Offloading in SDN-Based Fog Computing Systems. *Future Generation Computer Systems*, *117*, 486–497. 10.1016/j.future.2020.12.021

Pradhan, B., Hussain, M. W., Srivastava, G., Debbarma, M. K., Barik, R. K., & Lin, J. C. (2022). A Neuro-Evolutionary Approach for Software Defined Wireless Network Traffic Classification. *IET Communications*. 10.1049/cmu2.12548

Sharma, P. K., & Park, J. H. (2021). Blockchain-Based Secure Mist Computing Network Architecture for Intelligent Transportation Systems. *IEEE Transactions on Intelligent Transportation Systems*, *22*(8), 5168–5177. 10.1109/tits.2020.3040989

Tang, X., Cao, C., Wang, Y., Zhang, S., Liu, Y., Li, M., & He, T. (2021). Computing Power Network: The Architecture of Convergence of Computing and Networking Towards 6G Requirement. *China Communications*, *18*(2), 175–185. 10.23919/jcc.2021.02.011

Chapter 10

Cooperative network paradigm for device-centric nodes

Abraham George
Alliance University, Bangalore, India

10.1 INTRODUCTION

Future generations of wireless networks are expected to migrate to cooperative network architectures from the current connection-oriented architectures. Firstly, it is important to understand the characteristics of connection-oriented architectures to understand the differences between the two. In connection-oriented architectures, the user or the device negotiates for all service needs with a configured network attachment point. This can be termed a greedy approach, where the user demands for resources for its applications and the network attachment point will grant the resources to a user based on a fair policy of resource distribution. In cooperative network architectures, devices share resources allocated to them with other devices to achieve better overall performance. An example of cooperative communication is a relay node that transmits data on behalf of a neighboring node using resources allocated to it. Many times the application needs vary with respect to time and resource requirements may not be constant for a time period. The primary difference between the two architectures is that in the latter the device priority is to negotiate with the peer devices to collectively meet its need. In the case of cooperative networks all devices in the network request their resources based on the applications running at the device. Resources are primarily the spectrum, or the physical layer resources required to transmit/receive data over the air. In traditional cellular networks, it is assumed that all service requests will have the same characteristics but in futuristic applications such as augmented reality (AR)/virtual reality (VR)/metaverse/gaming, and others. It is expected that an application will comprise a group of networked devices with sensors and wireless interfaces. In this scenario, the resource needs will vary depending on the device type, application, quality of service (QoS), and the physical layer channel characteristics. The objective of every device is to transmit the data via the best available physical path to satisfy the service quality of the overall application. Wireless spectrum is a scarce resource which is the medium for data transmission.

DOI: 10.1201/9781003369028-10

Forthcoming wireless generations are exploring novel methods and mechanisms of packing more data onto the network. Terahertz communication, beam forming, and massive MIMO are all forthcoming technologies to yield higher data rates. While these technologies are being explored, it is important to explore directions to utilize existing spectrum and radio resources more efficiently. Wireless network cooperation is a concept where nodes share their resources to support services. Cooperative wireless communication refers to a network paradigm where wireless devices collaborate and cooperate with each other to improve overall system performance (Hossain et al., 2011). It leverages the capabilities of multiple devices working together to overcome the challenges posed by wireless channels and enhance the efficiency, reliability, and coverage of wireless communication systems.

The idea of wireless network cooperation initially came into existence in a scenario where a single antenna system in a multiuser scenario can share their antenna or spectrum, creating a virtual MIMO system (Nosratinia et al., 2004). In cooperative mode, a user will receive and transmit packets meant for itself and as well act as a forwarding agent for packets meant for neighboring nodes. This concept primarily increases the spectrum utilization, where a channel allocated for a particular node is utilized to transmit data to other nodes in vicinity of that node. The primary idea is that single-antenna mobiles in a multi-user scenario can share their antennas in a manner that creates a virtual MIMO system. A key aspect of this paradigm is to form a group of nodes that can work in a cooperative manner. These groups can be virtual, and distance may not be the only criterion for group formation. The goal of this chapter is to further illuminate the subject by broadening the scope of cooperation to other layers of the network, and thus accelerate the pace of developments in this exciting technology.

10.2 TYPES OF WIRELESS NODE COOPERATION

In traditional wireless communication systems, devices typically operate independently and establish direct links with a base station or access point. However, cooperative wireless communication takes a different approach by enabling devices to work together as a group or team, as illustrated in Figure 10.1. In Figure 10.1, the individual user nodes are both a relay and the user. This collaboration can occur between devices of the same type, such as mobile phones or laptops, or even between different types of devices, including sensors, relays, and access points. The fundamental concept behind cooperative wireless communication is to combine the resources of multiple devices to achieve mutual benefits. This collaboration can take various forms, including cooperative relaying, cooperative beamforming, and cooperative sensing.

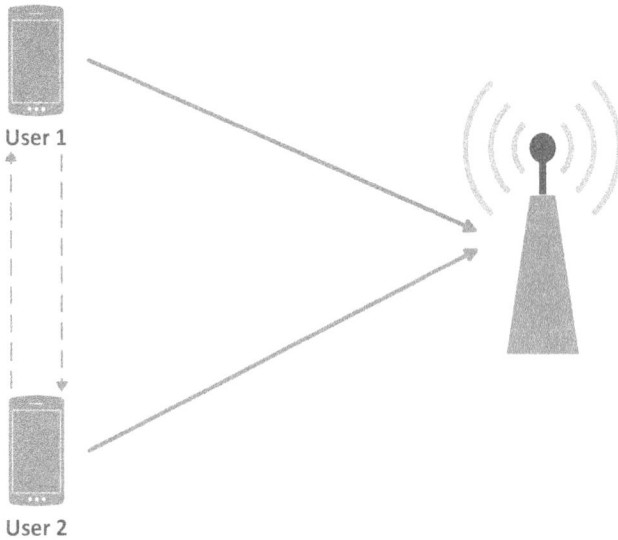

Figure 10.1 Cooperative communication.

10.2.1 Cooperative relaying

In cooperative relaying, devices cooperate to transmit and receive signals for each other. When a direct link between a source and a destination is weak or obstructed, intermediate devices act as relays to help forward the information. These relays receive the signal from the source, amplify it, and retransmit it to the destination. By exploiting spatial diversity and the availability of multiple relays, cooperative relaying can enhance signal quality, extend coverage, and mitigate fading and interference effects.

10.2.2 Cooperative beamforming

Beamforming involves shaping the radiation pattern of antennas to concentrate the transmitted power in a specific direction. In cooperative beamforming, devices collaborate to create coherent and robust beams. By combining their individual antennas and aligning their beamforming weights, devices can transmit signals with increased power and directivity towards the intended receiver, improving the signal quality and reducing interference. Cooperative beamforming is particularly useful in multi-user scenarios and dense wireless networks.

10.2.3 Cooperative sensing

Cooperative sensing involves multiple devices working together to gather and share information about the wireless environment. Devices can cooperatively detect and estimate parameters such as channel conditions, interference levels, or the presence of other devices. By combining their sensing capabilities, devices can make more accurate decisions, optimize resource allocation, and enhance spectrum efficiency.

Cooperative wireless communication offers several advantages. It can improve overall system capacity, coverage, and reliability by mitigating the effects of fading, interference, and obstacles (Cardona, 2022). It also enables the efficient use of network resources, such as bandwidth and power, by sharing them among cooperating devices. Moreover, cooperative techniques can enhance energy efficiency, prolonging the battery life of devices, and reducing energy consumption. Cooperative wireless communication finds applications in various domains, including cellular networks, ad hoc networks, sensor networks, and Internet of Things (IoT) systems. It has the potential to enable seamless connectivity, improve user experience, and support emerging technologies such as smart cities, autonomous vehicles, and industrial automation.

However, cooperative wireless communication also poses challenges in terms of coordination, synchronization, and the overhead of communication between devices. Efficient cooperation mechanisms, protocols, and algorithms need to be designed to ensure effective collaboration among devices. Furthermore, security and privacy considerations should be addressed to protect the integrity and confidentiality of shared information. Overall, cooperative wireless communication is a promising approach to overcome the limitations of traditional wireless systems. By leveraging the power of collaboration, it enables more efficient, reliable, and robust wireless communication networks, paving the way for advanced wireless applications and services.

10.3 SCENARIOS FOR NODE COOPERATION

Wireless network evolutions are driven by user scenarios. For example, 4G was driven by the need for higher data rates and 5G was driven by high user density and further higher data rates. Further network generations, 6G and beyond, are driven by emerging scenarios such as smart city sensor deployments and augmented reality (AR)/virtual reality (VR) applications. A smart city is a technologically advanced urban area that uses sensors to measure and monitor various data to improve quality of life for its citizens and streamline urban services. Large cities across the world have initiated projects to make their areas more citizen friendly in terms of transportation and utility services, waste management, energy consumption, and hospital services. There have been studies on the level of intelligence that is to be embodied in urban areas to qualify as a smart city (Halegoua, 2020). From a technological standpoint, the relevant aspect is to create a technological architecture that can scale to various levels of intelligence. The Internet of Things (IoT) is the prime enabler of smart cities that enables acquiring various kinds of data and performing an action. Urban areas around the world have created IoT deployment strategies based on the services it wants to offer to citizens and problems it intends to resolve (Alablani & Alenazi, 2020). The Internet of Things (IoT) is a group of inter-related or interconnected computing devices performing a mechanical, electric, or biological function. An IoT network can also be referred to as a wireless sensor network (WSN). A WSN can be broadly defined as embedded devices that communicate wireless in an ad-hoc or infrastructure less network. In a WSN, the connectivity with the public network is via a gateway node. Therefore, IoT devices can be termed as devices with direct or in direct connectivity with the public network.

The other significant application driving future network generations is the metaverse. Ever since Facebook announced its foray into the immersive technology with the metaverse as the broad theme, there has been serious speculation and thought into how this technology will change the world as we know. Even prior to Facebook's renaming to Meta and official announcement, there has been applications particularly in gaming using the Oculus lens and gaming platforms such as Roblox. Nvidia has been pursuing immersive technology since 2017 through the omniverse. Researchers and educators have been exploring and outlining the potential applications of virtual reality and augmented reality in gaming and people networking spaces even before the term *metaverse* caught everyone's attention. The significant newbie to the concept of VR is a digital avatar, which makes a digital replica of a human or a product. Digital avatar opens plethora of possibilities and in fact presents a radical change in how we currently experience human to human and human to machine interactions. These possibilities have triggered the interest of researchers in industry and academia anticipating revolutionary changes in many domains primarily manufacturing, education, and healthcare. The metaverse will be a set of immersive, interconnected digital spaces representing the evolution of the internet as we know it today. For example, in the metaverse, a digital replica of the product like an automobile can be tested and experienced by users. In healthcare, patient consulting and advanced diagnostics can be virtualized, providing users human experiences remotely. A digital twin (Haag & Anderl, 2018) is one of the core intrinsic themes of the metaverse. A digital twin is a virtual model of an object or system generated using real-world data for the purpose of understanding about its real-world counterpart and how it would function under various circumstances. In the case of the metaverse in healthcare, the digital twin could be the patient themselves, where doctors can analyze how the patient responds to a surgery and reaction to specific medical procedures or medicines.

Metaverse end-user equipment is a collection of wireless devices that will include several sensors, glasses, and headsets where data is ported

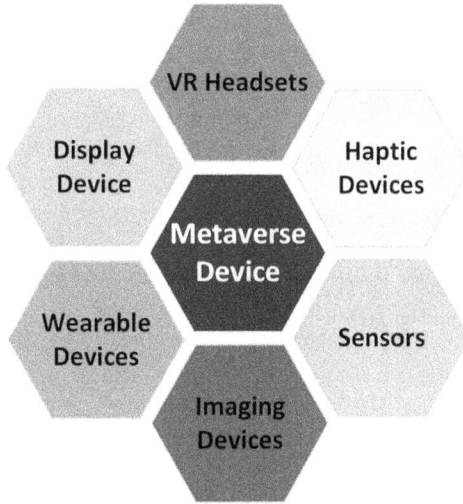

Figure 10.2 Form factor of a metaverse device.

amongst these devices or from the devices to a group central node and from there to a cloud node, fog node, or any other computing node, as depicted in Figure 10.2. The type of application data generated from and to these devices includes high-resolution images, video content, and large amounts of data both in structured and unstructured form (Ball, 2022). Currently, data generated from a region is proportional to the number of people in the region, assuming a certain number of devices per person and certain amount of data generated from each device. In the metaverse scenario, the notion of a device is loose or is not bounded by physical proximity as several disparate devices, human interface devices work in unison to provide the metaverse experience. In other words, a single metaverse user system comprises several devices that might share data among themselves or send to a master unit that collates the user data. This implies that there is data transfer or collation among devices and a greater amount of data load on the networks. These scenarios call of significant change or innovation in multiple directions as these forth-coming scenarios will generate massive load on the network and single technology may be largely insufficient to handle these loads. In this context, we introduce and extend the concept of cooperative networks, which is mostly limited to the physical layer in current literature and research (Hong et al., 2010).

The concept of cooperative networks was first introduced in cellular networks for enhancing the signal quality and capacity improvement. This concept resides predominantly at the physical layer. The goal of this chapter is aimed at extending the concept of cooperation and sharing to the upper layers of the networks where the concept of dynamic formation of network is illustrated. This chapter narrows down to the topic of node cooperation for wireless sensor networks. Node cooperation is discussed in cellular networks to improve network performance where nodes work collectively together to improve spectrum efficiency. The goal of this chapter is to highlight node cooperation as an effective and efficient mechanism to address network communication, and network management challenges. The term *cooperative communication* implies assistance from peer or neighbouring network nodes or collective intelligence. Collective intelligence implies that a group of nodes share information with each other where the decision making is based on inputs from all the participating nodes. This is a recent paradigm in computation which will find high applications in high-density sensor networks. The paradigm of cooperation can be extended beyond the lower layers of channel utilization to increase efficiency and capacity of the system. This has implication on node deployment strategies. The focus of this chapter is to highlight and explore the paradigm of network cooperation to increase the capacity of wireless networks. Future wireless networks are expected to become more device-centric where device has the capability to dynamically select and change the network attachment points to meet the application needs. Consider the case of a metaverse device that comprises haptic sensors, eyeglasses, and mobile devices. The devices will need to communicate with each other for a metaverse application. These devices may be individually performing an application as well. In this scenario, an application may communicate with the peer via a path that will meet its service needs. This will imply that a device may attach or associate itself with multiple attachment points to service its applications. This is the concept of device-centric nodes, where a device makes an effort to service its multiple applications that may have varying needs in terms of data rates, latency, and quality of experience (QoE). This will require collective intelligence of the nodes in the network where nodes cooperate with each other to assist each other. This chapter elaborates on the wireless node cooperation aspect for large-scale deployment that encompasses WSNs and metaverse scenarios. We will refer to these futuristic network scenarios as future networks (FNs).

10.4 DEVICE COOPERATION

The central theme of this chapter is to further outline and extend the concept of device cooperation at the higher layers in large-scale deployments particularly for smart city scenarios and metaverse scenarios. Device prior sections sufficiently set the context and the premise to arrive at this section. The limited resources of the FN nodes make the design of a large-scale sensor network and metaverse applications challenging (Silva et al., 2017). FN nodes will be deployed in smart city and metaverse scenarios where power is a constraint and there are stringent requirements on the application QoS. Application requirements, in terms of latency, data throughput, or device lifetime, often conflict with the network capacity and energy resources. One approach for addressing these trade-offs in literature is to rely on wake-up scheduling MAC protocols, which consists of alternating the active and sleep states of the nodes. In the active state, all the components of a node (CPU, sensors, radio) are active, allowing the node to collect, process, and communicate information. FN nodes generate large amounts of data and this data in large FN networks needs to be routed across the nodes, which considerably impact the latency and energy efficiency of the communication. There has been considerable literature to show that the performance of a WSN network can be further improved, if nodes not only synchronize, but also desynchronize with one another (Liu, 2015). Desynchronization refers to where nodes on different branches of the routing tree are active at different times to minimize or avoid congestion. In a FN, nodes can be logically organized in groups. In other words, the activity schedules of nodes that need to communicate with one another are synchronized to improve the message throughput. These nodes belong to a group or cohort and can be visualized as formation of network groups on the fly.

The system is intelligent, such that the schedules of the group of nodes that do not need to communicate are desynchronized to avoid congestion and packet losses. The cooperation of multiple FN groups may be able to improve the network lifetime of each FN node by load balancing all over the FN nodes because the energy consumption of nodes is almost largely dominated by data communication rather than by sensing and processing. Hence, the whole network lifetime can be prolonged by balancing the communication load at heavily loaded nodes around a sink. There are numerous studies on load balancing using clustering approaches in large WSNs (Shahraki et al., 2020). The concept of virtual or dynamic group formation results in creating localized application at the higher network layers as well. For example, consider a scenario where sensors are deployed in a dense urban area to monitor vehicular traffic density and road vibrations. The same devices will have multiple sensors, but the granularity of the measurement is controlled or created at the service discovery and the cooperative layers. In this chapter, the goal is to extend the concept of localized group formation beyond the sensor network applications for which there is multiple literature.

A group of devices will have to work cohesively to yield the required output. The optimum number of nodes required to yield desirable measurable accuracy for an application depends on spatial and temporal variation of the measurement. Measurements such as temperature will have gradual spatial variation and higher temporal variations. Hence, the groups must be formed dynamically based on the application characteristics and same node may be partisan in multiple groups to yield the desirable results. The term *cooperative networks for WSNs* refers to the ability or capability of the

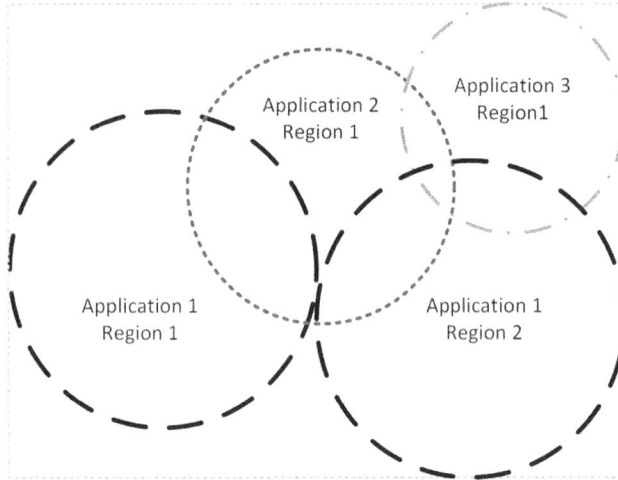

Figure 10.3 Sensor nodes forming regions through cooperation.

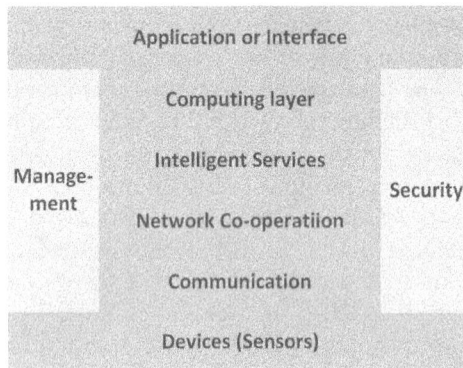

Figure 10.4 Layer of intelligence in device-centric nodes.

WSN nodes to inherently form virtual groups to conserve the available resources such as device power and frequency resources. Moreover, this approach also reduces the communication latency and drop rates, thus improving the reliability of communication.

Figure 10.4 illustrates this concept or theme. This is a scenario where nodes are connected to each other to reach the gateway node.

Figure 10.3 illustrates the theme of cooperation in large-scale networks where dynamic regions are formed and unformed based on node movement and application characteristics. In Figure 10.4, the rectangular box represents a given area. There are several co-located sensors in the region and several IoT applications are deployed in the region. Zones are dynamically formed regions based on several parameters such as node density, measurement granularity, type of measurement, network conditions, and power source constraints. Nodes in a zone may be connected to the public network directly or via a gateway node. In Figure 10.2, 'Application 1 Region1' and 'Application 1 Region 2' are non-overlapping regions making the same measurement while 'Application 2' and 'Application 3' regions overlap with Application 1 and region size is not uniform. The central idea of forming localized regions is to obtain higher performance and conserve limited resources. The scheme of zoning further results in node-cooperation whereby nodes interact with few essential neighbours to perform their task and conserve their resources. There exist several literatures (Shahraki et al., 2020) where this theme is used to maximize a specific metric such as power resources, data protection, routing efficiency, application performance, and others. There are multiple protocols and frameworks created for IoT scenarios (Sinche et al., 2020). The first step in the design of a large deployment scenario is to highlight the characteristics of the network that will subsequently enable the narrowing down of the protocols and frameworks at each layer. The next step is to explore strategies to fulfill the characteristics and maximize the efficiency of the network. The concept of zoning is primarily aimed at created localized regions. The larger research challenge is to develop strategies that operate dynamically because many a times sensor nodes are deployed in batches and in many cases the devices are not static. Examples of non-static sensor devices are devices mounted on moving vehicles and machines. Bio-inspired nature-inspired algorithms such as swarm optimization algorithms are effective in dynamic group formation (Singh et al., 2021). Swarm optimization algorithms are dynamically responsive, and the computation is distributed. An effective deployment strategy is highly dependent on the application and deployment scenario. Node cooperation is an effective strategy to build highly scalable IoT scenarios. These strategies combined with other strategies such as edge and cloud computing will be enabled to create large-scale IoT networks (Singh & Mohan, 2019). The objective of this chapter is to highlight node cooperation as an effective strategy for smart city IoT development. It is beyond the scope of this chapter to explore algorithms and approaches on each metric. Once smaller groups are formed, several metrics can be optimized through this approach. The key open areas with respect to collaboration are summarized in Table 10.1.

The primary contribution of this chapter is to extend the concept of node cooperation beyond the physical layer. To realize this concept, there must be multiple layers of intelligence, as illustrated in Figure 10.4.

The computing layers and services layer illustrated in Figure 10.4 continuously evaluate the network performance and enable the device to form virtual groups and attach to the gateway that best meet the application needs.

Table 10.1 Summary of challenges

Challenge	Description
Node auto-configuration	Devices should get auto-configured to the sensor network and the localized group.
Group formation	Dynamic formation of groups and rebalancing of groups within the sensor networks to maximize the performance.
Virtual group formation	Co-located or overlapping groups can exist to increase performance. A device can be part of multiple virtual groups to meet different application needs.
Creating sensor networks at scale	Network can accommodate nodes with varying underlying protocols.
Information passing across layers	Information should flow across layers for the computational layers to identify the attachment points. This should be dynamic as network parameters are highly time variant.

10.5 CONCLUSION

In this chapter, we highlight and elucidate the concept of node cooperation. This concept is highly relevant for smart city and metaverse applications. The early part of this chapter walks through the definitions and concepts of futuristic networks leading to the functional blocks of a cooperative device. Node cooperation is the key theme of this chapter, which is discussed in multiple sections. The goal here is to sufficiently emphasize the concept of node cooperation and its relevance, particularly in futuristic deployments. This should enable technology implementers to be aware of methods to improve the efficiency of their deployments and enable new research directions.

REFERENCES

Alablani, I., & Alenazi, M. (2020). EDTD-SC: An IoT Sensor Deployment Strategy for Smart Cities. *Sensors*, *20*(24), 7191. 10.3390/s20247191
Ball, M. (2022). *METAVERSE: And How It Will Revolutionize Everything*. Liveright Publishing Corp.
Cardona, N. (2022). *Cooperative Radio Communications for Green Smart Environments*. CRC Press.
Haag, S., & Anderl, R. (2018). Digital Twin – Proof of Concept. *Manufacturing Letters*, *15*, 64–66. 10.1016/j.mfglet.2018.02.006
Halegoua, G. R. (2020). *Smart Cities*. MIT Press.
Hong, P., Huang, W.-J., Kuo, J., & Springerlink (Online Service). (2010). *Cooperative Communications and Networking: Technologies and System Design*. Springer.

Hossain, E., Kim, D. I., & Bhargava, V. K. (2011). *Cooperative Cellular Wireless Networks*. Cambridge University Press.

Liu, X. (2015). Atypical Hierarchical Routing Protocols for Wireless Sensor Networks: A Review. *IEEE Sensors Journal, 15*(10), 5372–5383. 10.1109/jsen.2015.2445796

Nosratinia, A., Hunter, T. E., & Hedayat, A. (2004). Cooperative Communication in Wireless Networks. *IEEE Communications Magazine, 42*(10), 74–80. 10.1109/mcom.2004.1341264

Shahraki, A., Taherkordi, A., Haugen, Ø., & Eliassen, F. (2020). Clustering Objectives in Wireless Sensor Networks: A Survey and Research Direction Analysis. *Computer Networks, 180*, 107376. 10.1016/j.comnet.2020.107376

Silva, B. M. C., Rodrigues, J. J. P. C., Kumar, N., & Han, G. (2017). Cooperative Strategies for Challenged Networks and Applications: A Survey. *IEEE Systems Journal, 11*(4), 2749–2760. 10.1109/jsyst.2015.2436927

Sinche, S., Raposo, D., Armando, N., Rodrigues, A., Boavida, F., Pereira, V., & Silva, J. S. (2020). A Survey of IoT Management Protocols and Frameworks. *IEEE Communications Surveys & Tutorials, 22*(2), 1168–1190. 10.1109/comst.2019.2943087

Singh, A., Sharma, S., & Singh, J. (2021). Nature-Inspired Algorithms for Wireless Sensor Networks: A Comprehensive Survey. *Computer Science Review, 39*, 100342. 10.1016/j.cosrev.2020.100342

Singh, S., & Mohan Sharma, R. (2019). *Handbook of Research on the IoT, Cloud Computing, and Wireless Network Optimization*. IGI Global.

Chapter 11

Edge computing and edge intelligence

Srikanth Itapu[1], G. Ramana Murthy[1], and Mohan Krishna[2]
[1]Department of ECE, Alliance University, Bangalore, India
[2]Department of EEE, Alliance University, Bangalore, India

11.1 INTRODUCTION

With billions of mobile devices and billions more of data streaming worldwide, a disruption in the field of wireless communications has taken center stage. Technologies such as 5G/6G and beyond are bound to create new paradigms and opportunities in large-scale digitisation cutting across developing and developed nations. At the cusp of physical infrastructure, edge-based service-oriented architectures have gained traction in recent years. This culminates in the cloudification of novel network architectures that transform traditional networking to edge-cloud servers in the vicinity of mobile subscribers. This is the essence of edge computing (EC). With the emergence and interlinking of the Internet of Things (IoT), more and more data are being added by widespread and geographically distributed mobile and IoT devices. Ericsson predicted that approximately 40–45% of the 50 ZB data will be attributed to IoT devices by the year 2024. Offloading large chunks of data from the edge to the cloud might result in network destabilization and bottlenecks. This chapter covers the use case of computational offloading, emphasizing the research into novel architectures as a layered method for decentralizing edge computing.

The idea of edge intelligence (EI) extends to acquiring, storing, and processing data. The data processing is located at the edge (between a sensor and an IoT core) and storage services are in the cloud. The application of EI in smart appliances, wearables, industrial machines, automotive driver assistance systems, and smart buildings continues to increase with the integration of microcontrollers and embedded connections. By introducing intelligence to the edge computing nodes (ECNs), systems become

1. more decisive by efficiently adapting to the machine learning (ML) algorithms to reduce the frequency of contacting cloud servers.
2. Adhere to data security and follow local regulations.

DOI: 10.1201/9781003369028-11

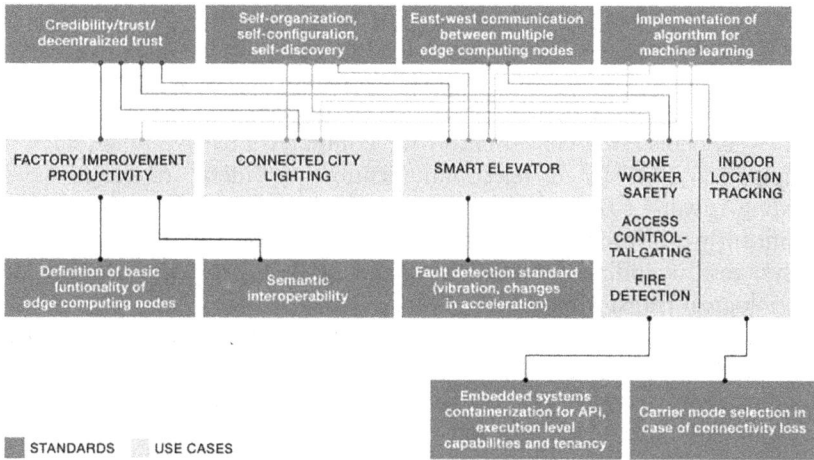

Figure 11.1 Edge intelligence (EI) standards with a few use cases.

3. Greatly reduce communication costs by using cache algorithms to pre-process data.
4. Adapt to temporary failures and manage maintenance procedures.

Depending on numerous use cases, some standard needs to be established for EI to address the challenges of ever-growing data dependencies. Figure 11.1 shows some of the standards needed for EI. Technical challenges that need to be considered are:

1. Self-configuration
2. Semantic interoperability
3. Fault detection standards
4. Credibility and trust

The cutting-edge uses of edge technology that are the main market drivers are highlighted in this chapter, such as smart manufacturing, building edge computing/edge intelligence architectures, and using cybertwin for off-loading in multi-access edge computing (MEC) for 6G applications.

11.2 LITERATURE SURVEY

In recent years, the MEC paradigm has attracted great interest from both academia and industry researchers. As the world becomes more connected, 5G promises significant computing, storage, and network performance advances in different use cases. This is how 5G, in combination with AI, has the potential to power large-scale AI applications, for example, in agriculture or logistics. The new generation of AI applications produces a huge amount of data and requires a variety of services, accelerating the need for extreme network capabilities in terms of high bandwidth, ultra-low latency, and resource consumption for computer-intensive tasks, such as computer vision. Hence, telecommunication providers are progressively trending toward MEC technology to improve the provided services and significantly increase cost-efficiency. As a result, telecommunication, and IT ecosystems, including infrastructure and service providers, are in full technological transformation.

Full-fledged intensive reviews on MEC were conducted and presented by authors (Ksentini et al. 2020), determining EI/EC framework, performance, and comprehensive overview of the state of the art, challenges, and further research directions. Heterogenous architectures were proposed by Yang et al. (2019), which empowered MEC with HetNets. An overview of leveraging MEC technology to achieve IoT applications and synergies between them was discussed in Corcoran et al. (2016). Introducing layback architecture to facilitate access to resources in software-defined networking (SDN) management for MEC was demonstrated by Hu et al. (2015) and Al-Ansi et al. (2021).

To disseminate the recent advances in edge intelligence, Sun et al. (2019) have conducted a comprehensive and concrete survey of the recent research efforts on edge intelligence. They survey the architectures, enabling technologies, systems, and frameworks from the perspective of AI models' training and inference. Some works also study the concept from the perspective of AI-driven fog computing (Mao et al. 2018; Peng et al. 2016). For example, Peng and Zhang comprehensively summarized the recent advances in performance analysis and radio resource allocation in the fog-radio access networks (F-RANs). This survey presents the advanced edge cache and adaptive model selection schemes to improve spectral efficiency (SE) and EE (Zhou et al. 2019). The authors also survey the F-RANs from the perspectives of the system architecture and key techniques, where the latter includes transmission mode selection and interference suppression (Peng et al. 2016). In addition, Mao et al. studied state-of-the-art research on the applications of deep learning algorithms for different network layers (Deng et al. 2020).

The confluence of AI and edge computing is natural and inevitable. In effect, there is an interactive relationship between them (Murali Kashaboina 2021). On the one hand, AI provides edge computing with technologies and methods, and edge computing can unleash its potential and scalability with AI; on the other hand, edge computing provides AI with scenarios and platforms, and AI can expand its applicability with edge computing. Figure 11.2 presents the road map for research in EI.

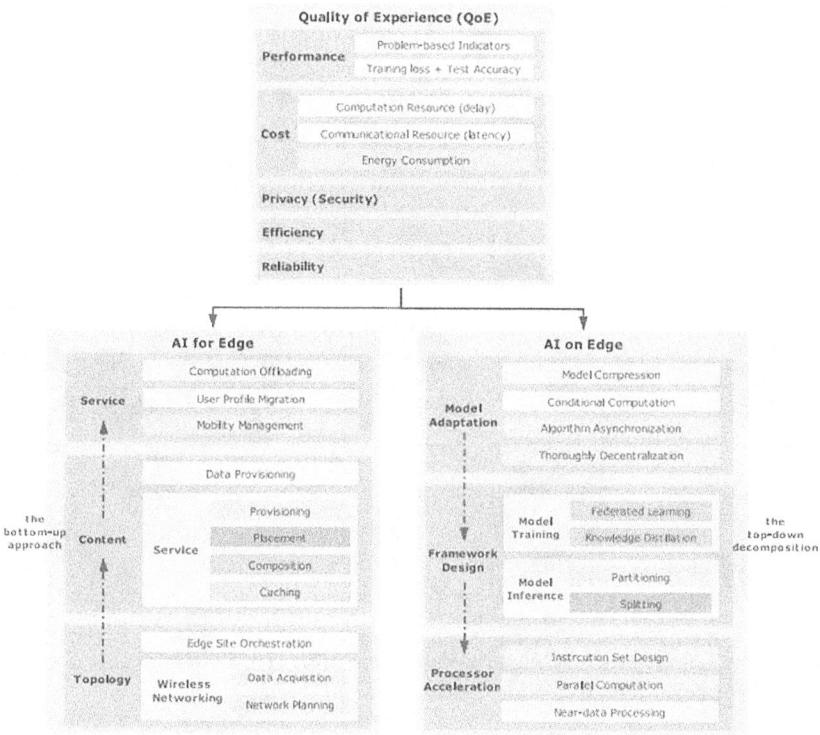

Figure 11.2 Road map of edge intelligence.

11.3 EMERGING ARCHITECTURE

Several experts in the industry proposed to develop a reliable and strong reference architecture for edge computing. While the goal of architecture development is to bring powerful computing to the edge, several architectures still can't decentralize edge computing sufficiently off the cloud. Here, a reference architecture that takes a layered approach to decentralize edge computing and address numerous known challenges is proposed. Figure 11.3 represents the proposed reference architecture.

The architecture has three distinct layers: device, edge, and cloud. The edge layer is central to the reference architecture that addresses edge computing requirements. The following are the key responsibilities of the edge layer:

- receiving, processing, and forwarding data flow from the device layer;
- provide time-sensitive services, such as edge security and privacy protection;
- edge data analysis;

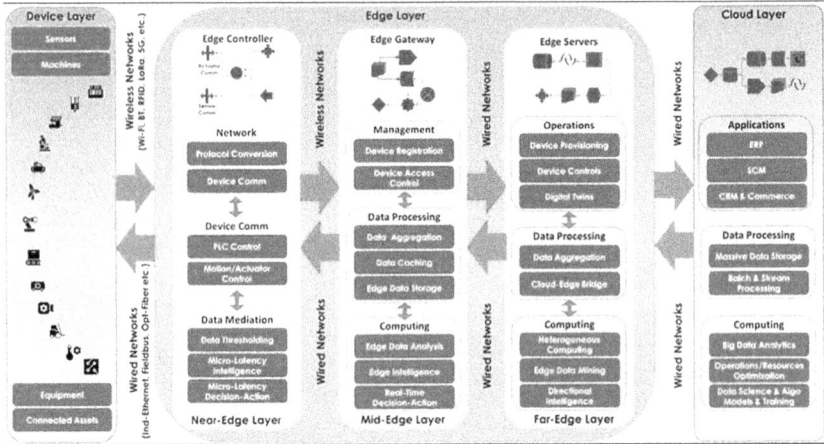

Figure 11.3 The architecture outlines the components at the device, edge, and cloud layers.

- intelligent computing; and
- IoT process optimization and real-time control.

The edge layer has three sublayers according to their data processing capacity: near-edge, mid-edge, and far-edge layers.

11.3.1 Near-edge layer

The near-edge layer contains edge controllers that collect data from the device layer, perform preliminary data thresholding or filtering, and control flow down to the devices. Because of gadget heterogeneity in the device layer, the edge controllers in the near-edge layer must support a wide array of communication protocols. The edge controllers also interface with upper layers to receive operational instructions or data-driven decisions and translate them into programmable logic controllers or action module-based control flow instructions to be transmitted to the devices. As a result, the near-edge layer must exhibit microsecond latency while interfacing with the device layer. Such latency becomes mandatory in situations where the call to action is time critical, such as the expected transient response of a self-driving vehicle in the event of a pedestrian suddenly entering the field of vision.

11.3.2 Mid-edge layer

The mid-edge layer contains edge gateways and is primarily responsible for exchanging data with the near-edge and the far-edge layers through wired and wireless networks. This layer has more storage and computing resources compared to the near-edge layer. More involved data processing can occur at this layer by combining information from multiple devices. The expected latency in this layer is milliseconds to seconds. Because this layer has storage capability, data and intelligence derived from the data processing can be cached locally to support future processing. The edge gateway in the mid-edge is also responsible for transferring control flow from the upper layers to the near-edge layer and managing the equipment in both mid- and near-edge layers.

11.3.3 Far-edge layer

The far-edge layer contains powerful edge servers responsible for performing more complex and critical data processing and making directional decisions based on the data collected from the mid-edge layer. Essentially, the edge servers in the far-edge layer form a mini-computing platform with more powerful storage and computing resources. The far-edge layer processes bulk data by using more complex machine-learning algorithms. The layer analyzes more data from different equipment to achieve process optimization or evaluates the best measures to take over a wider area for an extended period, usually with longer latency. The far-edge layer also acts as a bridge between the cloud and the edge layers.

11.4 HARDWARE EVOLUTION IN EC/EI

11.4.1 Data center evolution

Data centers are at the center of modern software technology, serving a critical role in the expanding capabilities of enterprises. The evolution of the data center has passed through three stages: siloed data centers, virtualized data centers, and software-defined data centers. The siloed data center relies heavily on hardware and physical servers, networks, and storage. It is defined by the physical infrastructure, which is dedicated to a singular purpose and determines the amount of data that can be stored and handled by the data center. This results in very low asset utilization at the price of high operational and capital expenditure. It can take an enterprise a month to deploy new applications with a traditional data center.

As technology is always pushing forward, resources are being pooled to create larger, more-flexible, centralized pools of computing, storage, and networking resources. A major challenge in IT today is that organizations can easily spend 70% to 80% of their budgets on operations, including optimizing, maintaining, and manipulating the environment. Likewise, how to handle dynamic workloads poses a significant challenge. Software-defined data centers combining server virtualization, software-defined networks (SDN), software-defined storage (SDS), and automation will enable the creation of a truly dynamic, virtualized data center. The control is exerted totally by software. That includes automating control of deployment, provisioning, configuring, and operation with software, creating one centralized hub for monitoring and managing a network of data centers. The software also exerts automated control over the physical and hardware components of the data center, including power resources and the cooling infrastructure, in addition to the networking infrastructure. Worth noting in this context is the trend within data centers to shift to smaller, more agile (i.e., movable) data centers towards the edge. This includes, for example, "edge caching" approaches, adopted by companies such as Google to minimize latency in response, using scaled-down "out of the box" data centers, co-located or situated near to ISP nodes. This can be taken to its logical conclusion if such caching/hosting is implemented at the base station level in wireless networks. Finally, this trend can be extended to allow containerization at the base station and thus enable the docking of third-party applications there.

11.5 IoT GATEWAYS/EDGE SERVERS

As billions of end devices need to connect to the world, one of the most critical components of future IoT systems may be a device known as an IoT gateway. Traditional gateways have mostly performed protocol translation and device management functions. They were not intelligent, programmable devices that could perform in-depth and complex processing on IoT data. Today's smart IoT gateways are full-fledged computing platforms running modern operating systems (OSs), such as Linux and Windows. New-generation IoT gateways are opening huge opportunities to push processing closer to the edge, improving responsiveness and supporting new operating models. The IoT gateway serves as an important bridge between operations and IT and provides a cost-effective business model. By adding IoT gateways, the current field deployment could require no change to run new applications such as predictive maintenance on the gateway. In scenarios such as smart manufacturing, with more and more robots, computer numerical control (CNC) machine tools and the like are generating massive real-time data in the field, and greater computing and storage resources will be necessary. In such situations, a local cloud at the edge represents a good choice. More edge servers will be interconnected and provide pooled and scalable resources. Several software programs or services can be deployed on an edge node simultaneously. To fulfill these demands, the ECN should have enhanced computing, storage, and networking capabilities, which require more powerful hardware. Firstly, the central processing units (CPUs) on edge nodes should have greater power, higher frequency, and larger L1 and L2 caches. Secondly, the edge nodes should have more extensive random-access memory (RAM) and flash memories. Thirdly, in some cases, the edge nodes should be more capable of dealing with harsh environments, e.g., vibration, wide temperature variations, dust, and electromagnetic interference in industrial sites and outdoor environments.

11.6 SMART SENSORS/END NODES

In addition to the development of end devices, higher-end sensor devices have become substantial computing devices in their own right. It is an emerging trend that this device class is becoming sufficiently powerful not to require a gateway device, or indeed through application docking to function as a gateway themselves, and thereby reduce the role of a gateway purely to that of a firewall/security function.

Illustrating that a relatively simple device can also be a powerful computing tool, this device, depicted in Figure 11.4, is at its base a radio frequency identification (RFID) proximity card reader but additionally has several inputs and outputs (four inputs and two outputs), as well as RS485 and Wiegand inputs, and can establish its connection via internet protocol (IP) with a remote server and channel, not only for its local data (and cache it for independent operation-based on business rules) but also manage other devices of a lower order connected to it, including voice over IP (VoIP) communication. Therefore, it performs multiple functions of the local sensor, edge server, and gateway. The terminal also includes local application docking, which can be used to manage and gather data from other devices connected to its peripherals, as well as operate identity-based applications, such as scheduling and biometrics.

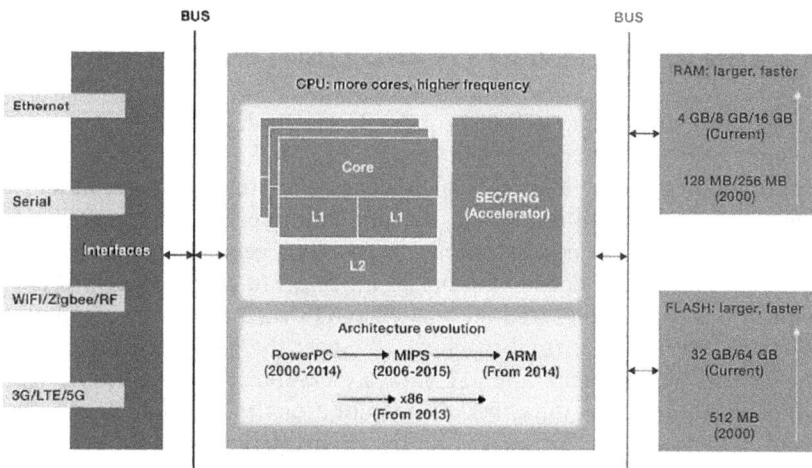

Figure 11.4 Hardware diagram of ECN.

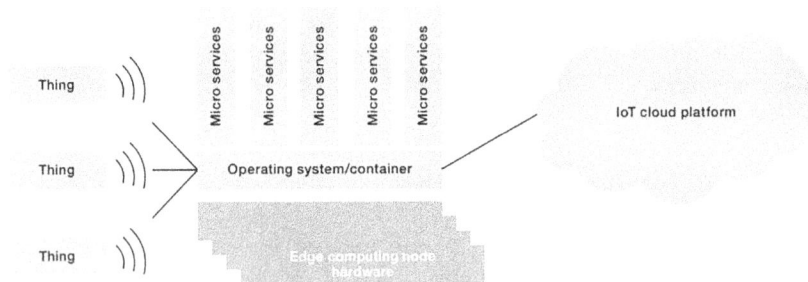

Figure 11.5 Software evolution and edge computing.

11.7 SOFTWARE EVOLUTION

11.7.1 IoT edge computing

Most of the IoT ECN technology, i.e., the technology, which is the host for edge processing capabilities, runs on flavors of the Linux OS while using different kinds of processor architectures, as shown in Figure 11.5. An industry-wide trend is emerging to package edge computing capabilities into microservices and deploy them within containers on IoT ECNs. For energy-efficient IoT infrastructures, the end devices have to hold local policies concerning when to sense or when to connect to the network. For example, in the case of environmental parameters monitoring, the important sensing runs during peak traffic hours when pollution might become a threat to public health. To convey device-specific capabilities, e.g., connectivity intervals, new IoT-tailored device management was standardized, called OMA Lightweight M2M (LWM2M).

11.8 ECN IOT GATEWAY OS AND LIGHTWEIGHT OS

In edge computing, data will be processed, analyzed, and aggregated at the network edge near things or data sources. The IoT gateway is the perfect host of all these capabilities. The main edge computing functions of an IoT gateway include cloud offloading, private data filtering, data aggregation, etc. The OS running on IoT gateways is usually a general-purpose OS such as Linux. Horizontal decoupling brings openness to gateways. Third-party applications can be deployed on a gateway OS via a Host OS, a container (LXC, Docker), or a virtual machine. To meet the high performance and low latency requirements, the IoT gateway OS integrates various network protocols to general Linux, makes ameliorations to the data forwarding plane, and selectively introduces hardware acceleration to some applications such as encryption/decryption. It is necessary to guarantee the security at the OS level, including RAM and storage security, the secure operating

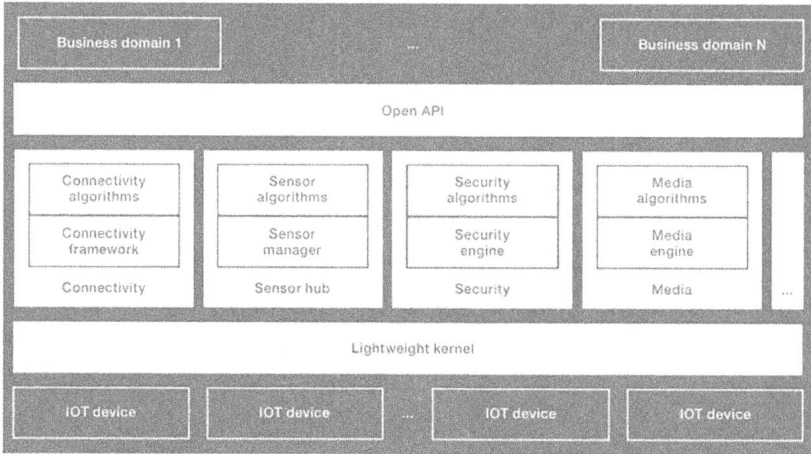

Figure 11.6 Lightweight OS architecture.

environment of the host OS, containers and virtual machines, the security of encryption keys, anti-injection of malware, secure operating environment, and malfunction isolation of third-party applications, etc.

Lightweight OS kernel running on system-on-chips in tiny devices will hide the chipset difference between silicon companies, and this OS provides the drivers and reacts to events happening around the hardware, as shown in Figure 11.6. Considering the resource-constrained applications, the kernel must have a light footprint, high start-up speed, low power consumption, and fast response. The lightweight OS can be referred to as a class of OSs including Tiny OS, Contiki, LiteOS, Mantis, etc. The point of the open API is to abstract the chaotic world of a system on a chip (SoC) design and the complexity of key services (such as connectivity) away from developers, leaving a cleaner, common interface to work with. Programmers who are handy with C++, JavaScript, HTML, Swift, and other languages for phones, tablets, and desktops can prototype and build applications for fiddly hardware ultimately hidden away under the platform. These programmers do not need to know about the undocumented registers and control bits and the rivalries plaguing the system-on-chip world. This is abstracted away by the open API, which tries to pave over the fragmentation.

Today's IoT largely exists in isolation, and it has been impossible to realize a truly interconnected world where devices are noninteroperable. The import of the lightweight OS platform to the vertical EI will open a new chapter. Since the platform provides consistent APIs over the connectivity, security, application, and other such domains, which hides the vendor's implementation differences, it is evident that it will fundamentally solve the interoperability issues of terminals (sensors), and terminals (sensors) with applications.

11.9 CURRENT STATE-OF-THE-ART IN EDGE INTELLIGENCE

Like many other IoT-enabling technologies, however, machine intelligence (MI) research and development has largely been restricted to the IT sector, as the complexity of convolutional neural networks (CNNs), hidden Markov models (HMMs), natural language processing, and other disciplines used in the creation of ML algorithms and deep neural networks (DNNs) require storage and computing resources usually only accessible on a data center scale. One of MI's early excursions into the OT space came with the release of the NVIDIA Jetson TK1 platform in 2014. Based on the Tegra K1 SoC and its 192-core Kepler graphics processing unit (GPU) and quad-core ARM Cortex-A15, the Jetson TK1 brought data center-level to compute performance to computer vision, robotics, and automotive applications, but also provided embedded engineers with a development platform for the CUDA deep neural network (cuDNN) library. The cuDNN primitives' enabled operations such as activation functions, forward and backward convolution, normalization, and tensor transformations required for DNN training and inferencing, and the combination of this technology with the Jetson TK1's 10 W power envelope meant that deep learning frameworks such as Caffe and Torch could be accessed and executed on smaller OT devices. In addition to high-cost embedded approaches, pattern-matching low-level instructions in embedded processors have been added (e.g., Intel Quark).

11.10 EDGE COMPUTING ARCHITECTURE FOR INDUSTRY

11.10.1 Industrial Internet Consortium architecture

The industrial internet is an Internet of Things, machines, computers, and people, enabling intelligent industrial operations using advanced data analytics for transformational business outcomes. The Industrial Internet Consortium (IIC) reference architecture, following the OMG/ODP tradition and the IEC guidelines, considers the four viewpoints: 1 v) business, 2 v) usage, 3 v) functional, and 4 v) implementation. The functional viewpoint in turn considers five domains: 1d) control, 2d) operations, 3d) information, 4d) application, and 5d) business, supported by the six common security functions: audit, identity, cryptography, privacy, authentication, and physical protection. This three-tier model is one representative of the implementation view. Figure 11.7 describes the three-tier model, in which 1d is mainly at the edge tier, 2d and 3d are mainly at the platform tier, and 4d and 5d are at the enterprise tier. Consequently, EI requires migration of some of 3d and 4d to the edge domain. Intelligence, along with security, resilience, analytics, etc., is considered by IIC as a key system concern; it relates to the highest, i.e.,

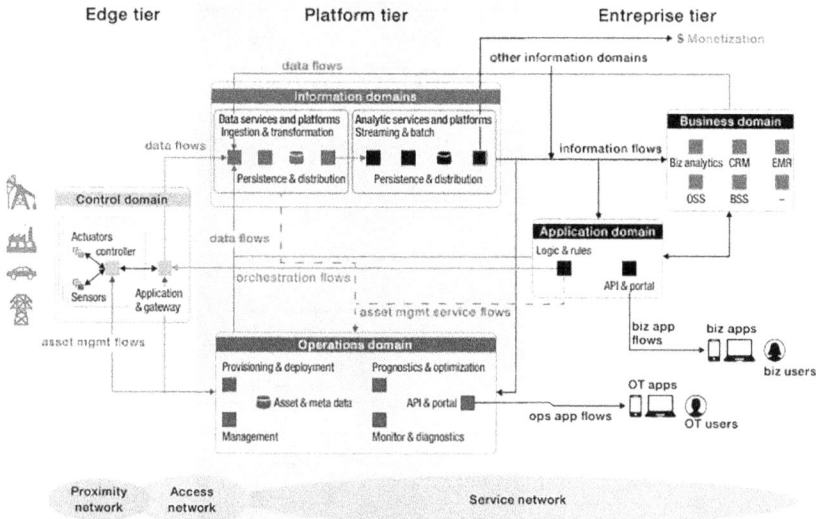

Figure 11.7 IIC architecture.

the third, level of understanding in communication, meaning that it facilitates the interpretation of the sender's intent. Intelligence is to be supported in several places; for example, intelligent decisions can be supported by automated service discovery. However, the key place-holder of the IIC intelligence is intelligent and resilient control (IRC). The IIC architecture document sketches several modelling considerations relevant to the IRC design, along with a sample IRC workflow, and maps them to the functional components such as planners, predictors, blame assigners, and ethical governors.

11.11 MULTI-ACCESS EDGE COMPUTING ARCHITECTURE

European Telecommunications Standards Institute (ETSI) multi-access edge computing (MEC) working group provides a new approach to mobile core networking. Operators can open their radio access network (RAN) edge and place authorized third-party application functionality towards mobile subscribers, enterprises, and vertical segments. The mobile edge computing framework shows the general entities involved in enabling the implementation of mobile edge applications as software-only entities. These can be grouped into system-level, host-level, and network-level entities as shown in Figure 11.8. Mobile edge management comprises the mobile edge system-level management and the mobile edge host-level management.

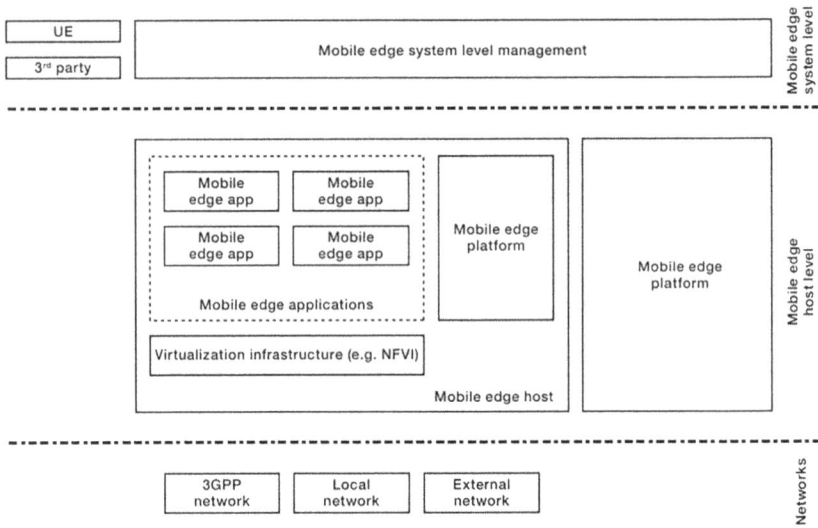

Figure 11.8 ETSI MEC framework architecture.

The architecture of the MEC server is composed of middleware services to the applications:

- Infrastructure services, including (1) communication services providing API to interact with the application services and between them, and (2) a service registry used by the applications to discover and locate the endpoints for the services they require.
- Radio network information services (RNIS) provide information to the applications to calculate and present the following high-level and meaningful data: cell ID, location of the subscriber, and cell load and throughput guidance.
- Traffic offload function (TOF) prioritizes traffic and routes the selected, policy-based, user data stream to and from applications that are authorized to receive the data. It can act in a pass-through mode, in which the data plane is sent to an application and then to the original PDN-GW, or in an endpoint mode by serving the data to the location application.

In September 2016, the industry specification group (ISG) MEC changed the acronym MEC to "multi-access edge computing" to reflect the importance of addressing Wi-Fi and fixed line in addition to 3GPP access technologies. Opening up the radio access networks to third-party players can create value and opportunities for mobile operators and accelerate innovation for new services and applications that make use of proximity, context, and speed available at the mobile edge. In 2021, the New Industry Specification Group on Reconfigurable Intelligent Surfaces (RIS) was initiated. ETSI DECT-2020 NR was approved by ITU-R as the first non-cellular 5G technology and the MEC sandbox for edge app developers were launched.

11.12 COMPUTATION OFFLOADING IN MEC

11.12.1 Application of cybertwin for 6G networks

Cybertwin (Rodrigues et al. 2021; Yu et al. 2019, 2020) is a recently designed new system for MEC. As such, the Cybertwin can be also seen as a sort of virtual server, built to be an avatar of the user in the MEC system. The Cybertwin was created as a replacement for VM servers. As such, it works between the MEC system and the users, receiving user tasks and giving them resources for their completion. This also means that under the Cybertwin system, cloud servers do not communicate with users, interfacing only with Cybertwin instead. Because cloud servers and users do not interact directly with each other, even if the user moves and due to this mobility ends up assuming a new network address (e.g., by connecting to a new access point), as long as the Cybertwin retain the same address, communication can continue with any problem. Moreover, this separation between user and server means that Cybertwin migration is easier than VM server migration, by just informing the user and the server of the new Cybertwin address in the network. Additionally, the task of sharing and keeping information is also easier in Cybertwin-based frameworks due to the introduction of a control plane. Cybertwins are built with the idea of frequently reporting the control plane regarding the usage of the network around them, the usage of the server where they are hosted, and information regarding the tasks they receive from their users. In exchange, the control plane can aggregate all this data and use this global knowledge of the system to determine optimal configurations, such as when to migrate Cybertwins and where tasks should be executed. These decisions are then shared with the Cybertwins. This, together with the capability that Cybertwins have of offloading tasks to any cloud servers (whereas VM servers are usually limited to the cloudlet where they are hosted (Rodrigues et al. 2018) makes it possible to reach optimal performance in the form of the fastest execution of tasks through a balanced workload between all cloud servers, a very important metric for QoE at MEC (Rodrigues et al. 2017; Alfakih et al. 2020). Figure 11.9 shows a simple representation of the Cybertwin-based MEC, with a focus on how the control plane aggregates information from all Cybertwins and, thus, allows for enhanced coordination and distribution of workload. Cybertwins have other important advantages, such as an activity log of important actions and the capability of negotiating the sale of information to interested companies if the user is

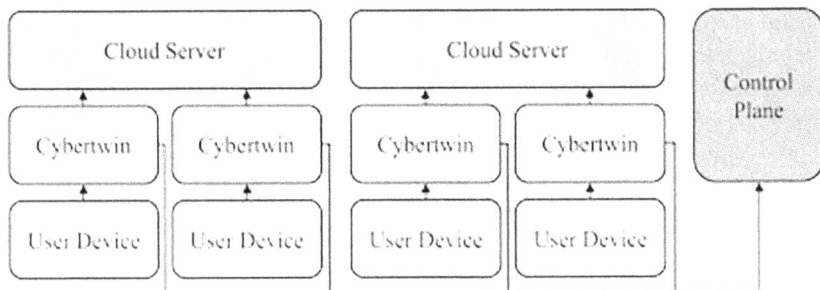

Figure 11.9 Cybertwin-based MEC.

interested in it. These points do not matter much for the study in this article, but readers interested in how Cybertwins work and these other benefits are referred to the bibliography. Measuring the extent to which the control plane can enhance MEC performance is the main objective of this chapter. This is important as Cybertwin has some challenges when being deployed for MEC. First, there is overhead in the communication between the Cybertwin and the control plane. Additionally, VM server-based systems are already established, meaning that a transition to Cybertwin is difficult and costly due to the need of changing all cloud-based systems to the new standard. This change is only worthwhile if Cybertwins can bring a significant performance improvement. Based on this, an assessment is made of how much the adoption of Cybertwin might reduce service delays.

11.13 TIME ALLOCATION POLICY IN WIRELESS POWERED MOBILE EDGE COMPUTING

The battery life of mobile terminals (MTs), time allocation policy for energy harvesting and data offloading, and offloading policy based on an optimal number of subtasks were investigated for a single MT in the WP-MEC system, as shown in Figure 11.10. The application for computational offloading is divided into an optimal number of subtasks and then, through offloading policy, the decision to follow either remote execution or local execution is made. A detailed cost function which jointly considers time delay and energy consumption for offloading policy is pertinent to EC. In addition, the cost function includes the current battery level of MT and thus prolongs the battery life of MT. Deep learning approaches are used to find the optimal number of subtasks per application and partial offloading policy along with time allocation policy for energy harvesting simultaneously. The deep learning–based offloading and time allocation policy (DOTP) algorithm provides a time allocation policy that results in extended battery life as compared to the benchmark schemes. A DNN is trained on a pre-generated data set through a mathematical model for the achievement

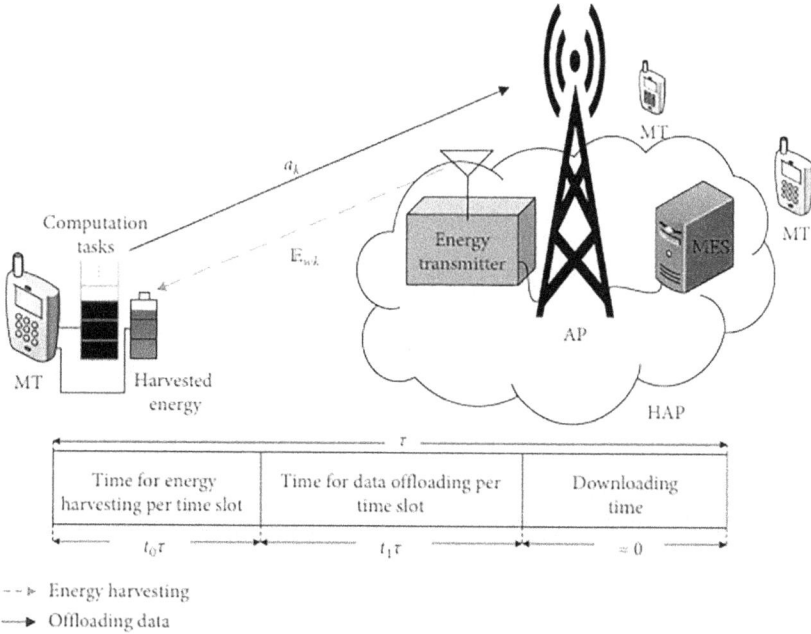

Figure 11.10 Wireless powered mobile edge computing system.

of objectives to reduce the energy consumption and time delay that incur for traditional optimization methods and make the decisions faster. It is evident from the numerical results that our proposed DOTP demonstrates extended battery life, lower energy consumption, and lower time delay for application execution in comparison with the benchmark techniques. In the future, the scenario can be extended to multiple MTs and HAPs, as well as to vehicular networks. In addition, a reinforcement approach can be used, and routing strategies can be investigated for handover management during mobility from one HAP to another (Numani et al. 2021).

11.14 INTELLIGENT REFLECTING SURFACES FOR MEC IN 6G NETWORKS

Wang et al. (2021) studied an IRS-empowered MEC adopting the non-orthogonal multiple access (NOMA) techniques. The energy efficiency is jointly optimized considering the offloading power, receiving beamforming, local computing frequency, and IRS phase shift. Zhou et al. (2021) considered IRS-assisted MEC architecture in which IRS is employed to support computation task offloading from users (two users) to an access node linked with an edge computing cloud. Chu et al. (2021) deployed and analyzed the computational performance of IRS-enhanced MEC architecture, as shown in

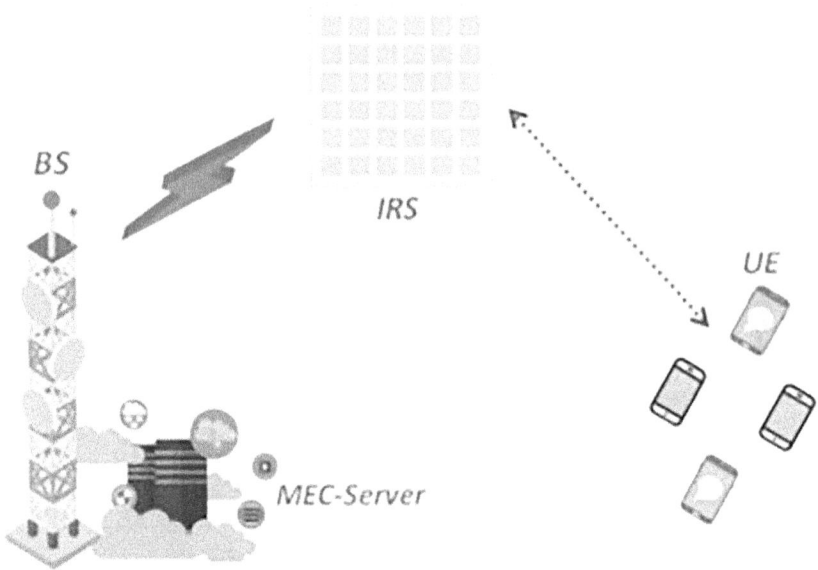

Figure 11.11 IRS-assisted communication scenario.

Figure 11.11. The work formulated a problem targeting optimizing sum computational bits considering the offloading time allocation, CPU frequency, transmit power of each device, etc. Zhang et al. (2021) investigated the network throughput optimization problem of an IRS-aided multiple-hop MEC system. The work jointly optimized resource allocation and phase shifts of the IRS. Bai et al. (2020) investigated the lucrative role of IRS in MEC architecture in which single-antenna user equipment may offload partial computational tasks or activities to the computing node (edge computing) via a multi-antenna access node with the assistance of an IRS. Latency-minimization problems are composed for both multi-device and single-device scenarios. Mao et al. (2021) aimed at utilizing the IRS to escalate the efficiency of wireless energy transfer and task offloading. Specifically, the work investigated the maximization of total computation bits for IRS-assisted wireless-powered MEC systems through the joint optimization of the transmission power, beamforming of IRS, and time slot assignment. Sun et al. (2022) presented a new IRS-MEC framework that jointly optimized the local computing frequencies (CPU cycles) of the user equipment (UE) and the offloading schedules to reduce the consumption of energy of the UE.

11.15 CONCLUSION AND FUTURE SCOPE

This chapter analyzed in depth the characteristics and benefits of edge computing (EC) and edge intelligence (EI), as well as the challenges, potential applications, and market drivers. These are a few conclusions:

- Integration of EC/EI in 6G technologies enables sufficient and massive support for different services, including IoT, smart environments, augmented and virtual reality, sustainability of energy systems, vehicle connection, network performance, video games, economic services, education technology, ICT markets, and communication.
- EI/EC's main characteristics and benefits include proximity of end users, ultra-low latency, location awareness, integrated virtualization, super network performance, flexibility, real-time analytics, and automation.
- EI/EC's current and up-to-date challenges include privacy and security, latency, distributed resource management, data traffic and bandwidth, heterogeneity, and scalability in addition to some issues related to network openness, multiple services and processes, data management, durability, and resilience.
- IEC's potential use cases and applications that have been investigated in this paper include three main categories: customer-oriented services, operator and third-party services, and network performance and QoE improvement. To enhance these categorizations, it is essential to integrate EI/EC in 6G technologies.
- EI/EC's market drivers vary based on the different use and accessibility of 6G technologies in various sectors. The use of AR and VR in the healthcare industry, autonomous vehicles, smart energy, and smart environments are the five main sectors that are highlighted.

REFERENCES

Al-Ansi, A., Al-Ansi, A. M., Muthanna, A., Elgendy, I. A., & Koucheryavy, A. (2021). Survey on intelligence edge computing in 6G: Characteristics, challenges, potential use cases, and market drivers. *Future Internet*, *13*(5), 118. 10.3390/fi13050118

Alfakih, T., Hassan, M. M., Gumaei, A., Savaglio, C., & Fortino, G. (2020). Task offloading and resource allocation for mobile edge computing by deep reinforcement learning based on SARSA. *IEEE Access: Practical Innovations, Open Solutions*, *8*, 54074–54084. 10.1109/access.2020.2981434

Bai, T., Pan, C., Deng, Y., Elkashlan, M., Nallanathan, A., & Hanzo, L. (2020). Latency minimization for intelligent reflecting surface-aided mobile edge computing. *IEEE Journal on Selected Areas in Communications*, *38*(11), 2666–2682. 10.1109/jsac.2020.3007035

Chu, Z., Xiao, P., Shojafar, M., Mi, D., Mao, J., & Hao, W. (2021). Intelligent reflecting surface assisted mobile edge computing for Internet of Things. *IEEE Wireless Communications Letters*, *10*(3), 619–623. 10.1109/lwc.2020.3040607

Corcoran, P., & Datta, S. K. (2016). Mobile-edge computing and the Internet of Things for consumers: Extending cloud computing and services to the edge of the network. *IEEE Consumer Electronics Magazine*, *5*(4), 73–74. 10.1109/mce.2016.2590099

Deng, S., Zhao, H., Fang, W., Yin, J., Dustdar, S., & Zomaya, A. Y. (2020). Edge intelligence: The confluence of edge computing and artificial intelligence. *IEEE Internet of Things Journal, 7*(8), 7457–7469. 10.1109/jiot.2020.2984887

Hu, Y. C., Patel, M., Sabella, D., Sprecher, N., & Young, V. (2015). *Mobile edge computing—A key technology towards 5G.*

Kashaboina, M. (2021, September 15). *An intelligent edge: A game changer for IoT.* IoT Agenda; TechTarget. https://www.techtarget.com/iotagenda/post/An-intelligent-edge-A-game-changer-for-IoT

Ksentini, A., & Frangoudis, P. A. (2020). Toward slicing-enabled multi-access edge computing in 5G. *IEEE Network, 34*(2), 99–105. 10.1109/mnet.001.1900261

Mao, Q., Hu, F., & Hao, Q. (2018). Deep learning for intelligent wireless networks: A comprehensive survey. *IEEE Communications Surveys & Tutorials, 20*(4), 2595–2621. 10.1109/comst.2018.2846401

Mao, S., Zhang, N., Liu, L., Wu, J., Dong, M., Ota, K., Liu, T., & Wu, D. (2021). Computation rate maximization for intelligent reflecting surface enhanced wireless powered mobile edge computing networks. *IEEE Transactions on Vehicular Technology, 70*(10), 10820–10831. 10.1109/tvt.2021.3105270

(N.d.-b). Etsi.org. Retrieved April 4, 2023, from https://portal.etsi.org/Portals/0/TBpages/MEC/Docs/Mobile-edge_Computing_-_Introductory_

Numani, A., Ali, Z., Abbas, Z. H., Abbas, G., Baker, T., & Al-Jumeily, D. (2021). Smart application division and time allocation policy for computational offloading in wireless powered mobile edge computing. *Mobile Information Systems, 2021,* 1–13. 10.1155/2021/9993946

Peng, M., Yan, S., Zhang, K., & Wang, C. (2016). Fog-computing-based radio access networks: issues and challenges. *IEEE Network, 30*(4), 46–53. 10.1109/mnet.2016.7513863

Rodrigues, T. G., Suto, K., Nishiyama, H., & Kato, N. (2017). Hybrid method for minimizing service delay in edge cloud computing through VM migration and transmission power control. *IEEE Transactions on Computers. Institute of Electrical and Electronics Engineers, 66*(5), 810–819. 10.1109/tc.2016.2620469

Rodrigues, T. G., Suto, K., Nishiyama, H., Kato, N., & Temma, K. (2018). Cloudlets activation scheme for scalable mobile edge computing with transmission power control and virtual machine migration. *IEEE Transactions on Computers. Institute of Electrical and Electronics Engineers, 67*(9), 1287–1300. 10.1109/tc.2018.2818144

Rodrigues, T. K., Liu, J., & Kato, N. (2021). Application of cybertwin for offloading in mobile multiaccess edge computing for 6G networks. *IEEE Internet of Things Journal, 8*(22), 16231–16242. 10.1109/jiot.2021.3095308

Sun, C., Ni, W., Bu, Z., & Wang, X. (2022). Energy minimization for intelligent reflecting surface-assisted mobile edge computing. *IEEE Transactions on Wireless Communications, 21*(8), 6329–6344. 10.1109/twc.2022.3148296

Sun, Y., Peng, M., Zhou, Y., Huang, Y., & Mao, S. (2019). Application of machine learning in wireless networks: Key techniques and open issues. *IEEE Communications Surveys & Tutorials, 21*(4), 3072–3108. 10.1109/comst.2019.2924243

Wang, Q., Zhou, F., Hu, H., & Hu, R. Q. (2021). Energy-efficient design for IRS-assisted MEC networks with NOMA. *2021 13th International Conference on Wireless Communications and Signal Processing (WCSP).*

Yang, X., Yu, X., Huang, H., & Zhu, H. (2019). Energy efficiency based joint computation offloading and resource allocation in multi-access MEC systems. *IEEE Access: Practical Innovations, Open Solutions*, 7, 117054–117062. 10.1109/access.2019.2936435

Yu, Q., Ren, J., Fu, Y., Li, Y., & Zhang, W. (2019). Cybertwin: An origin of next generation network architecture. *IEEE Wireless Communications*, 26(6), 111–117. 10.1109/mwc.001.1900184

Yu, Q., Ren, J., Zhou, H., & Zhang, W. (2020). A cybertwin based network architecture for 6G. *2020 2nd 6G Wireless Summit (6G SUMMIT)*.

Zhang, H., He, X., Wu, Q., & Dai, H. (2021). Spectral graph theory based resource allocation for IRS-assisted multi-hop edge computing. *IEEE INFOCOM 2021 - IEEE Conference on Computer Communications Workshops (INFOCOM WKSHPS)*.

Zhou, F., You, C., & Zhang, R. (2021). Delay-optimal scheduling for IRS-aided mobile edge computing. *IEEE Wireless Communications Letters*, 10(4), 740–744. 10.1109/lwc.2020.3042189

Zhou, Z., Chen, X., Li, E., Zeng, L., Luo, K., & Zhang, J. (2019). Edge intelligence: Paving the last mile of artificial intelligence with edge computing. *Proceedings of the IEEE Institute of Electrical and Electronics Engineers*, 107(8), 1738–1762. 10.1109/jproc.2019.2918951

Chapter 12

Network virtualization

Geetha A and Punam Kumari

Department of Computer Science and Engineering, Alliance University, Bangalore, India

12.1 INTRODUCTION

The UN SDGs for digital societies will be fulfilled through 6G's fusion of the physical, digital, and biological worlds. Further dramatic changes are necessary for 6G, and wireless transmission capabilities must be stretched to their absolute maximum. The massive use of AI in networks and applications, as well as radical developments, is required for future wireless business ecosystems. Embedded trust is necessary for a 6G network. Suitable attack mitigation and defense should be provided by the 6G network. Privacy protection and unambiguous market regulations will be essential facilitators in the new data marketplaces that 6G will establish. Moving from best effort to differentiated service quality requires an improved networking model for 6G. 6G is not just about transporting data; it will develop into a framework of services, including communication services (Dogra et al., 2020). All computation and intelligence that is specific to a certain user may be moved to the edge cloud in 6G. Many new applications for 6G are made possible by the combination of mobility, very accurate positioning, and sensing. A successful 6G service platform must have a strong foundation for trust and privacy.

The trajectory of cloud computing is being followed by network virtualization as well. Particularly cloud use of computing has made the transition from using local, invested physical resources to remote, virtual resources reachable over the popular internet. Network virtualization also encourages the transition of networking operations from physical gadgets nodes to a virtualized pool of resources. However, unlike cloud computing, network virtualization abstracts the network itself in addition to virtualizing computer and storage services. As a result, it may be more complicated to understand.

A continuum of virtual resources, from cloud computing to edge/fog computing up to smart ambient, interconnected by ultra-broadband and ultra-low latency links, will be used to replace the existing lack of truly end-to-end services; for example, in beyond-5G and 6G networks. Sustainability

DOI: 10.1201/9781003369028-12

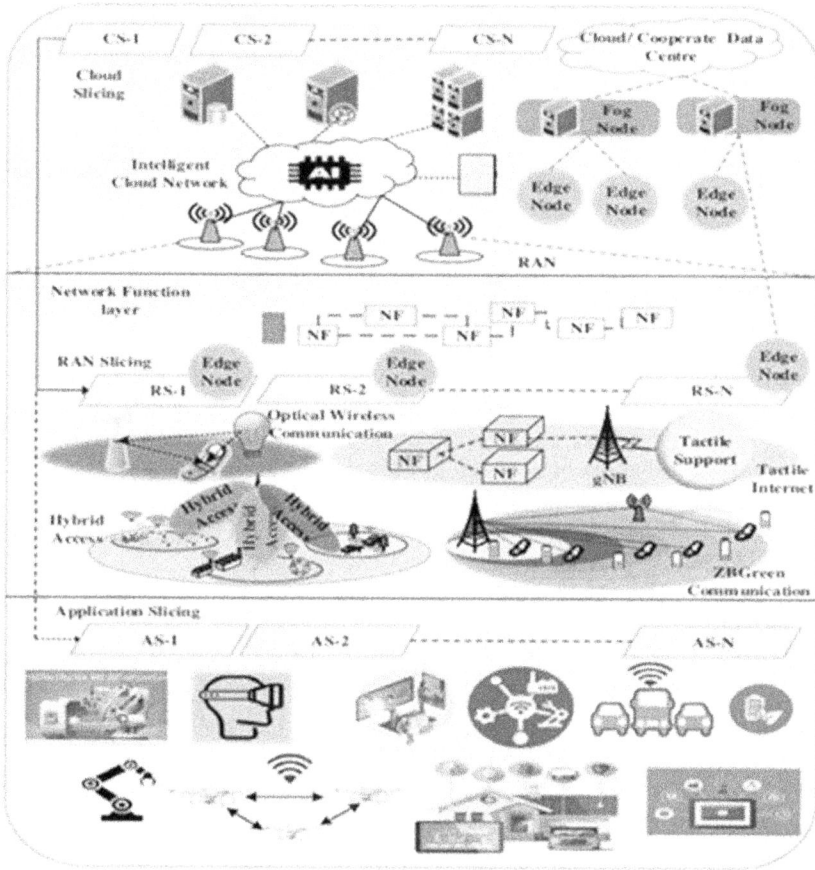

Figure 12.1 The virtualized network slicing architecture of a 6G network.

will necessitate multi-domain scenarios that involve numerous stakeholders and parties working together to provide end-to-end services in accordance with new business models such as network operators, service providers, and municipalities. A ubiquitous network telemetry system and AI that can adapt to the changing needs of cloud-native services and applications across the continuum of virtual resources will be used to take advantage of this dispersed orchestration and management (Figure 12.1).

This will mark a significant shift from the existing idea of "control what you observe" to a broad range of potential outcomes for the power plane and the operations, administration, and management (OAM) field. To support the transition to high precision services, networks beyond 5G and 6G would need to further reduce latency and improve reliability for precision-demanding services such as industrial internet services. This has been exceedingly challenging due to the unavailability of the necessary control automation and transport network technology. With edge computing and distributed architecture that offers more flexibility and support for network analytics, networks beyond 5G and 6G are anticipated to be fully cloud native. They will also advance more convergence with non-3GPP access.

12.2 NETWORK FUNCTION VIRTUALIZATION EVOLUTION

Knowing the industry's past and present challenges is necessary to comprehend the factors that led to the rapid adoption of NFV in networking. Data communication networks and hardware have developed throughout time. Despite their increasing speed and robustness, networks nevertheless struggle to fulfil the demands of the market. The networking industry is being driven by a new set of demands and challenges brought on by cloud-based services, such as infrastructure to support those services and expectations to make them run more efficiently. Examples include IoT applications, the quick increase of data-enabled devices, and sizable data centers with processing and storage.

12.2.1 Traditional network

Examples of the early data transfer networks include the conventional phone network and possibly even telegraph networks. Early on, latency, availability, throughput, and low data loss were the guiding principles and quality criteria for networks. These variables have a direct impact on the layout and features of the hardware and equipment used to transport the data. Physical devices also ran closely tied proprietary operating systems and only supported a small number of extremely specialized functions and use cases. The criteria and elements influencing network architecture and device performance did not change with the introduction of data transport networks.

The data networks that were created were specialized to effectively meet these efficiency goals because traditional networking equipment was made for certain jobs. It had a strong connection to it, was tightly integrated with the silicon field programmable and custom integrated circuits and was only intended to operate in the specific ways that the device required. These specialized hardware platforms were used to run this software or code. This situation forces the operators to come up with ideas on ways to remove the limits. Some of these limitations in a traditional network are high operational costs, concerns for migration, capacity over-provisioning, limitations in flexibility, scalability, time-to-market, manageability, and interoperability (Figure 12.2).

12.2.2 NFV introduction

Networks work virtualization strives to improve how network operators build networks by combining multiple types of network equipment and

Classical Network Appliance Approach

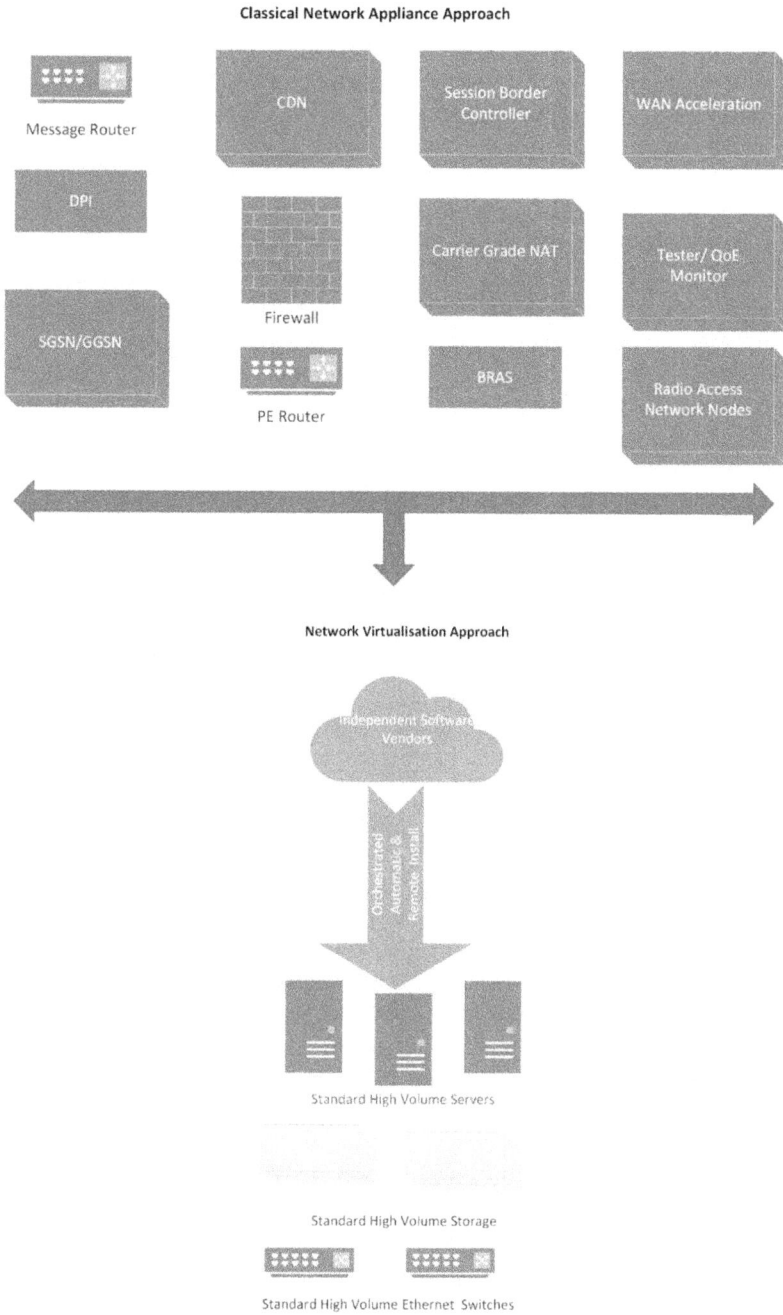

Message Router

DPI

SGSN/GGSN

CDN

Firewall

PE Router

Session Border
Controller

Carrier Grade NAT

BRAS

WAN Acceleration

Tester/ QoE
Monitor

Radio Access
Network Nodes

Network Virtualisation Approach

Independent Software
Vendors

Orchestrated
Automatic &
Remote Install

Standard High Volume Servers

Standard High Volume Storage

Standard High Volume Ethernet Switches

Figure 12.2 Vision for network function virtualization.

producing several servers, switches, and storage that can be exposed in data centers, network nodes, or client houses. Figure 12.1 serves as an example. These virtual appliances can be created whenever they are needed, without new hardware needing to be installed. Network administrators might, for instance, set up an open-source firewall on a virtual machine (VM).

12.3 NETWORK VIRTUALIZATION BACKGROUND

Networks work virtualization attempts to improve how network operators design networks by integrating distinct network equipment types by developing industry-standard numerous servers, switches, and storage that may be exposed in data centers, network nodes, or customer residences. As seen in Figure 12.1, these virtual appliances can be created whenever needed, without the installation of additional hardware. For instance, network administrators may use a VM to run an open-source firewall.

It calls for developing network services in software that can run on a variety of commercially available servers, relocate to or instantiate in different locations throughout the network as necessary, all without the need to set up additional hardware. To put it another way, network function virtualization encourages the integration of network functions into software that can be controlled (for example, transferred or replicated) without necessitating changes to the physical infrastructure and run on several widely used IT systems in data centers.

12.3.1 Network function virtualization

Network function virtualization, or NFV, is a concept in network architecture that uses virtualization technologies to separate hardware from network functions. The goal of NFV is to improve the scalability of communication service and computing services by virtualizing entire categories to deliver the new network node functions into modular units.

It enables the implementation of network activities using software that can be progressed to, or loaded into, different places throughout the network as necessary. This software can run on a kind of industry-standard server hardware. New hardware installation is no longer necessary as a result. The ETSI NFV is one of the most well-liked reference frameworks and architectural footprints for NFV.

12.3.2 Software-defined networking

Software-defined networking (SDN) pushes the intelligence already existing in network parts by merging network functions with software with the aid of a central controller. SDN supposedly centralizes the control plane of the network as opposed to older solutions that spread it across all network devices. By doing this, the deployment of new network features only requires software modification to the controller rather than significant, pricey equipment modifications or firmware updates. The flexibility this technique offers network management, enabling them to swiftly change their policy regarding how traffic is distributed over the network, is its main advantage.

The ONF OpenFlow Switch was fundamental for the original development of the protocol, which was designed exclusively for IEEE 802.1 switches. As the advantages of the SDN paradigm have spread to more people, its use has been expanded to increasingly tangled situations including wireless and mobile networks.

This protocol was originally created especially for IEEE 802.1 switches that adhered to the OpenFlow Switch. As more people become aware of the advantages of the SDN paradigm, more complicated scenarios, such as wireless and mobile networks, have been included in its scope of use.

12.3.3 Multi-access edge computing

When multi-access edge computing (MEC), also known as mobile edge computing, is used at the mobile network's edge, it can help to deliver services to mobile users in an effective and dynamic manner. The ETSI ISG MEC working group will start developing an open environment towards the end of 2014 to enable the integration of MEC capabilities with outside applications and service providers' networks. With the help of these distributed computing capabilities, mobile access networks can implement various functions using an IT architecture that is like that of a cloud environment. In this sense, it is obvious to an addition to NFV and SDN (OmniRAN).

IEEE 802.11CF proposes an access network that uses IEEE 802 standard-based technologies to connect terminals to their access routers such as 802.3 Ethernet, 802.11 Wi-Fi, etc. The standard creates an inter-entity communication testimonial for access networks that include entities, reference points, and behavioral and functional descriptions. By enabling shared-network control and the application, it aims to remove barriers that stand in the way of emerging network technologies, network operators, and service providers.

12.3.4 Distributed management task force

The DMTF is a body that develops industry standards and aims to make it easier to administer network-accessible technologies through open and collaborative efforts with selected technology companies. The DMTF is actively involved in the creation and adoption of interoperable management standards, supporting implementations that enable the management of numerous established and emerging technologies, such as the cloud, virtualization, networks, and infrastructure.

12.4 NETWORK VIRTUALIZATION CHALLENGES

The deployment, expansion, and operation of networks in the telecommunications industry are changing because of network virtualization. These new technologies will bring down overall fetch by working communication aid from specialized hardware in operators' cores to server farms spread throughout data centers. Also, network virtualization has a fundamental impact on the way networks connect the components of a network service route, process, and regulate traffic.

12.4.1 Network softwarization of SDN/NFV

SDN research focuses on both autonomous network administration and the optimization of traffic management to meet the demanding needs of next-generation networks. This includes load balancing, multi-path routing, service route quality management, and automatic route repair. It also includes dynamic real-time automated network management. Network function virtualization (NFV) divides the actual network into several independent virtual networks, each with its own operating resources, to offer various service quality alternatives. The security challenge is attacks on controllers and network service disabling. Network deployment flexibility rises along with device and operational cost reductions as NFV's significance as a research area develops. We can increase system performance by combining the best tele-traffic pathway management with SDN principles.

12.4.2 5G and network slicing

The fact that 5G will need to support a lot more use cases than previous wireless generations, which initially prioritized phone services and later voice and high-speed packet data services, has been established among the outset of every 5G discussion in the research areas and industry. In 5G, it should be able to support new services like the haptic internet and the Internet of Things in addition to the same (or better) phone and packet data services. (IoT). Some of these use cases demand extremely low latency and higher speeds, while others demand extremely low power consumption and excellent delay tolerance, pushing the specifications to their limits.

12.4.3 Device virtualization for end users

Virtualizing the functions of network element has been a focus of network softwarization strategies up to this point. Mobile network architectures are quickly switching over to virtualize radio access network operations as well, even though the network's core was the first to be virtualized. The next logical step is to bring virtualization all the way down to end-user devices, like virtualizing a smartphone.

Reproducing a mechanism in the cloud has appealing advantages for both network operations and the device, including power savings at the device due to offloading computationally intensive tasks to the cloud and improved networking between devices and infrastructure for service delivery thanks to tighter device integration.

12.4.4 Security and privacy

Security and privacy are two crucial considerations that must be made, just like in any other situation when resources are shared. It may be necessary for several service providers to cohabit in a virtual or hybrid environment when it comes to security. This requires confirmation procedures among diverse in effect and physical operations and resources in addition to ongoing external monitoring.

Like this, several network slices employed on the same framework can cause safety issues; for example, if a slice running a less important approach interrupts a slice running a vital request, such as support for a safety system. On a shared system, the minimum common denominator for security precautions should usually be equal to or higher than the one needed for the most crucial implementation. DevOps and numerous, ongoing threat model evaluations are necessary for an NFV system to maintain a specific level of security.3.5 Network Function Placement.

Any type of network telecommunications infrastructure has an issue with the placement of network functions. Furthermore, network virtualization's greater degree of freedom makes this issue more significant and challenging to solve. The location of VNFs is a resource allotment problem that may consider several factors, including resilience, anti-affinity, security, privacy, and energy efficiency. Placement algorithms grow more complicated and significant when numerous functions are connected.

12.5 NETWORK VIRTUALIZATION ARCHITECTURE

Network virtualization is the process of creating multiple virtual networks on top of a physical network infrastructure. This allows for greater flexibility and efficiency in network management and enables multiple tenants to share the same physical network resources. The architecture of network virtualization typically involves three layers.

12.5.1 Infrastructure layer

This layer consists of physical network infrastructure, such as switches, routers, and cables. The infrastructure layer provides the underlying network connectivity and enables the creation of virtual networks.

12.5.2 Control layer

The control layer is responsible for managing the virtual networks and their associated resources. This layer includes virtualization controllers and network orchestrators that manage the virtual networks and allocate resources to them. The control layer also provides a centralized point of control for network policies and security.

12.5.3 Application layer

The application layer consists of the applications and services that use virtual networks. This layer includes virtual machines, containers, and other network services that are hosted on the virtual networks.

12.6 NETWORK VIRTUALIZATION IN A 5G NETWORK

Network Slicing: It is one of the important features that enables the creation of multiple virtual networks, each with its own network policies, security measures, and quality of service (QoS) settings. These virtual networks, or slices, can be customized to meet the specific needs of different applications and services, and can be dynamically allocated and de-allocated as needed.

For example, a mobile network operator might create multiple network slices, each tailored to meet the specific needs of a different customer or application. One network slice might be optimized for low-latency, high-bandwidth applications such as virtual reality or gaming, while another might be optimized for mission-critical applications such as remote surgery or autonomous vehicles.

Virtualization using slicing enables network operators to effectively utilize their network resources, improve network performance, and better meet the diverse needs of their customers and applications. It also provides a more flexible and scalable network architecture, allowing network operators to quickly adapt to changing business needs and technological advancements.

12.7 NETWORK SLICING FOR VIRTUALIZATION

Network slicing is a technique to create several unique logical and virtualized networks on top of a multidomain infrastructure. Each logical network (slice) is created to support a specific application, users, or networks. Several techniques which are used to create slicing include the following (Figure 12.3):

1. Network Functions: They are the basic building blocks to create network slices. They are used to demonstrate elementary network functionalities.

Service Layer

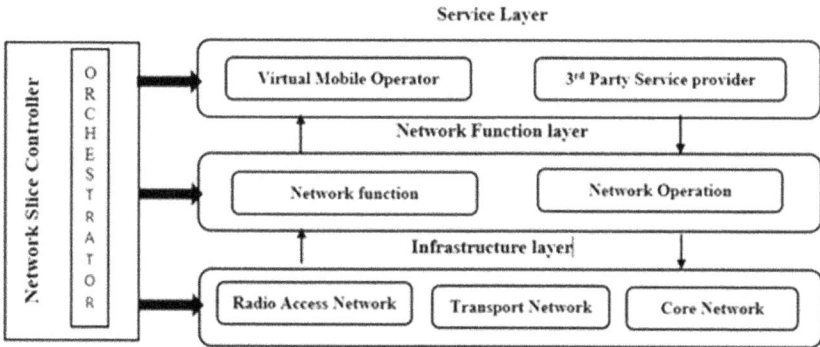

Figure 12.3 Generic network slicing framework.

2. Virtualization: It is used to provide an abstract view of all the physical resources under a unified scheme.
3. Orchestration: Here, the role of software defined networking (SDN) comes into the picture. Orchestration process is used to coordinate the working of all the network components that are involved in a complete life cycle of a network slice (Toosi et al., 2018).

The general network slicing framework can be composed of two main blocks. The first block contains actual slice implementation, and the second block works as controller to control and manage slices. Details of the blocks are given below.

12.7.1 First/primary block

The primary block contains all the functionalities which are related to slice implementation. It is a multi-tier architecture that consists of three layers: service, network, and infrastructure. Details of the layers are as follows.

12.7.2 Service layer

This layer works as an interface for various business entities.

12.7.3 Network function layer

This layer is responsible for creating network slices as per the requirement from the upper layer. Various network functions are combined and placed over a virtual network to create an end-to-end slice that is used to provide specific services demanded by the service layer. Here, network functions are nothing but the set of networks operations that are used to manage the complete life cycle of a slice. To increase resource utilization, more than one slice can share same network operation but the trade-off is that it will increase the operation management complexity.

12.7.4 Infrastructure layer

It is the actual physical topology on top of various slices of the networks that are multiplexed.

12.7.5 Second/controller block

As the name suggests, this block is used to control and manage all the functionality provided by various slices. It enables slices to reconfigure during their life cycle and provide effective coordination between the above defined layers. To reduce the complexity to manage several network operations performed by various slices, the controller block contained multiple orchestrators. These multiple orchestrators independently manage the subset of functionality of each layer.

12.8 VIRTUALIZATION IN A 6G NETWORK

6G networks are still in the early stages of development, and it is unclear exactly how they will be designed and deployed. However, it is likely that network virtualization will continue to play an important role in the evolution of mobile networks, including 6G.

Some of the potential ways that network virtualization could be used in 6G networks include the following.

12.8.1 Enhanced network slicing

6G networks are expected to provide even more advanced network slicing capabilities than 5G networks, allowing for even greater customization of virtual networks to meet the needs of different applications and services. A potential solution of this problem is AI-based network slicing in which AI-based algorithms play a significant role in various phases of network slicing (Kafle et al., 2018). A brief description about AI-based network slicing is given below.

12.8.1.1 Role of artificial intelligence in preparation phase

In the preparation phase, AI needs to perform two tasks:

1. Service demand prediction: Utilizing AI methods like recurrent neural networks, service demand may be forecasted based on past data. Prior research demonstrates that it is possible to anticipate with

accuracy the service demand and resource utilization of a slice. The outcomes of the forecast can be used to guide decisions during the planning stage (Hinton et al., 2015).

2. Slice admission: In order to maximize network resource utilization while taking resource availability and service needs into account, the SDN controller permits slices. This issue is classified as an integer optimization problem since the slice admission choice is binary. AI-based solutions have the potential to replace traditional optimization techniques in large-scale networks with complex resource availability distribution.

12.8.1.2 AI for planning

In the planning phase, AI can perform two tasks:

1. VNF placement: VNFs are introduced by the SDN controller to facilitate network services. To ensure service delay requirements, the resources allotted for VNFs should be dynamically modified for time-varying service demands. In dynamic network systems, resource utilization may be improved by using deep learning techniques.
2. Resource reservation: Depending on the needs of each slice, the SDN controller reserves resources. Since data traffic volumes fluctuate over time, resource reservations should be flexible enough to meet changing real-time needs. Reinforcement learning (RL) techniques like deep deterministic policy gradient (DDPG) can help with this.

12.8.1.3 AI for operation

Two exemplary operation tasks are as follows:

1. Resource orchestration: End users are given access to a slice's restricted resources. Based on real-time user mobility, service requests, etc., judgments are made. RL techniques may be used for dynamic resource orchestration to effectively use resources.
2. RAT selection: For each end user, the best RAT is chosen from a pool of many potential RATs in order to maximize system utility. User perception of a RAT's service performance is stochastic as a result of user mobility. Multi-armed bandit techniques, such as contextual bandit, can be used to solve this problem.

12.8.2 Edge computing and network function virtualization

Network virtualization can enable edge computing and network function virtualization in 6G networks, which can help reduce latency, improve network performance, and increase scalability. In order to support the performance, new features, and new services of future 6G networks, EI, powered by AI techniques (e.g., machine learning, deep neural networks, etc.), is already regarded as one of the key missing components in 5G networks (Xu et al., 2020). A unique type of cloud computing called edge computing (EC) transfers some of the processing and data storage related to a given service from the central cloud to edge network nodes that are situated physically and logically near to data suppliers and end users. Performance enhancements, traffic optimisation, and new ultra-low latency services are all features of current 5G networks. All of these features will benefit greatly from edge intelligence in 6G. Therefore, it can be predicted that as telecom infrastructures advance towards 6G, highly distributed AI would be taken into account, shifting the intelligence from the central cloud to edge-computing resources (Zhou et al., 2019). This indicates that the roles that various devices play are still largely predetermined at the time of their design and installation. Therefore, "liquid" software—software that can "flow" from one device to another—is still a long way off. Future 6G networks will be unable to move calculations without design-time partitions; therefore, without liquid software, we are forced to choose where to put the intelligence in the network architecture at the appropriate design time (Rausch & Dustdar, 2019).

12.8.3 Multi-cloud integration

Hosting data, apps, or infrastructure across many cloud providers is known as a multi-cloud strategy. Addressing common problems that multi-cloud integration typically raises include the following.

12.8.3.1 Architectural challenge

When moving solutions to the cloud or other cloud environments, an organization's current architecture must go through major changes, especially if they have common only employed on-premises technology. Organizations frequently need to rewrite their programs to run them in the new cloud environment.

12.8.3.2 On-premises integration structure maintenance

It can be challenging to maintain the integration structure of on-premises data and supporting systems when firms implement multi-cloud integration. If organizations wish to keep the current links between data and apps, they must compromise between the requirements of multi-cloud integration and those of conventional on-premise integrations.

12.8.3.3 Agility

When relying on a single cloud system or on-premises deployments, agility is less important than it is for multi-cloud integrations. For instance, having nodes in various cloud applications necessitates the capacity to swiftly move between these nodes at various times, which could make integrations more complex and cause delay.

12.8.3.4 Data protection

Under laws like the GDPR, organizations are responsible for their governance practises and policies relating to personally identifiable information (PII). All data processors, including cloud service providers, must adhere to the controller's instructions, which are effectively a contract in the form of traditional contractual clauses or legally binding corporate norms. With the use of multi-cloud solutions, these data processors become more prevalent and multi-cloud connections become more complicated and dangerous.

12.8.3.5 Containers and microservices

Microservice and containers are useful for cloud app deployment. Microservices and containers are both beneficial, but they are extra techniques that might need to be changed to work with any future cloud integration. These cloud resources can be smoothly managed and integrated with a variety of cloud environments that might be supported by 6G networks, thanks to network virtualization.

12.8.3.6 Network automation

Network virtualization in 6G networks can automate administrative procedures like resource management and QoS control. This could lead to higher productivity and decreased operational costs.

12.9 6G END-TO-END NETWORK AUTOMATION CHALLENGES

As was previously demonstrated, 6G networks are made up of a complicated infrastructure, a wide range of heterogeneous services, and dispersed use cases. Additionally, they have stricter performance standards. As a result, end-to-end network automation and orchestration for 6G systems becomes a crucial issue to take into account. To guarantee that these networks reach their full potential, a new set of difficulties related to 6G end-to-end network automation and orchestration must be examined. Some of them are given below:

1. Continuous Orchestration
2. Heterogeneous Orchestration
3. Multi-Stakeholder and Multi-Tenant Orchestration

4. AI-Driven Orchestration
5. Private Network Support
6. Advanced Monitoring
7. Security Architecture

Overall, network virtualization is expected to continue to be an important technology in the development of 6G networks, enabling greater flexibility, scalability, and efficiency in network management, as well as improved performance and enhanced security. To provide some research direction to the researcher, some open research opportunities in this domain are given below:

1. Dynamic Data-Driven Orchestration
2. 6G Traffic Prediction
3. Distributed Orchestration Decision Making
4. Orchestration Domain Adaptability

12.10 BENEFITS OF NETWORK VIRTUALIZATION

Network virtualization offers several benefits, including the following.

12.10.1 Improved resource utilization

With the help of network virtualization, one can make better use of their existing network resources, leading to cost savings and increased efficiency.

12.10.2 Simplified network management

Network virtualization provides a centralized point of control for network policies, security measures, and quality of service (QoS) settings. This makes it easier for network administrators to manage the network and ensure that all network resources are being utilized effectively.

12.10.3 Increased security

Virtual networks are isolated from each other, which means that security breaches on one virtual network are less likely to affect other virtual networks or the physical network infrastructure. This isolation provides an additional layer of security that can help protect sensitive data and applications.

12.10.4 Scalability

Network virtualization enables organizations to create new virtual networks quickly and easily as needed, without the need for additional physical network infrastructure. This makes it easier to scale the network to meet changing business needs.

12.10.5 Flexibility

Virtual networks can be customized with their own network policies, security measures, and QoS settings. This allows organizations to tailor the network to meet their specific needs and requirements.

12.11 CONCLUSION

Network virtualization is the practice of creating multiple virtual networks on top of a single physical network infrastructure. In the context of 6G, network virtualization is expected to play a significant role in enabling the next generation of wireless communications. 6G networks are expected to be highly heterogeneous, incorporating a variety of different technologies and architectures, including satellite communications, terahertz communications, and other forms of wireless networking.

By using network virtualization techniques, 6G networks will be able to dynamically allocate network resources to different applications and services, based on their specific requirements. This will enable the efficient use of network resources, while ensuring that applications and services receive the necessary quality of service (QoS) levels.

Additionally, network virtualization can help to increase network security by enabling the isolation of different network functions and services from each other. This can help to prevent the spread of security breaches or attacks across the entire network and can also enable more fine-grained control over network access and permissions. Overall, network virtualization is likely to be a key enabler of the 6G vision, enabling the creation of highly flexible and adaptable networks that can support a wide range of use cases and applications.

REFERENCES

Dogra, A., Jha, R. K., & Jain, S. (2020). A Survey on Beyond 5G Network with the Advent of 6G: Architecture and Emerging Technologies. *IEEE Access*, 9, 67512–67547. 10.1109/access.2020.3031234

Hinton, G., Vinyals, O., & Dean, J. (2015). Distilling the Knowledge in a Neural Network. arXiv preprint arXiv:1503.02531 [Cs, Stat]. https://arxiv.org/abs/1503.02531

Kafle, V. P., Fukushima, Y., Martinez-Julia, P., & Miyazawa, T. (2018). Consideration on Automation of 5G Network Slicing with Machine Learning. 2018 ITU Kaleidoscope: Machine Learning for a 5G Future (ITU K). 10.23919/itu-wt.2018.8597639

Rausch, T., & Dustdar, S. (2019, June 1). Edge Intelligence: The Convergence of Humans, Things, and AI. In *2019 IEEE International Conference on Cloud Engineering (IC2E)*, pp. 86–96. IEEE. 10.1109/IC2E.2019.00022

Toosi, A. N., Mahmud, R., Chi, Q., & Buyya, R. (2018). Management and Orchestration of Network Slices in 5G, Fog, Edge and Clouds. arXiv preprintarXiv:1812.00593 [Cs]. https://arxiv.org/abs/1812.00593

Xu, D., Li, T., Li, Y., Su, X., Tarkoma, S., Jiang, T., Crowcroft, J., & Hui, P. (2020). Edge Intelligence: Architectures, Challenges, and Applications. arXiv preprint arXiv:2003.12172 [Cs]. https://arxiv.org/abs/2003.12172

Zhou, Z., Chen, X., Li, E., Zeng, L., Luo, K., & Zhang, J. (2019). Edge Intelligence: Paving the Last Mile of Artificial Intelligence With Edge Computing. *Proceedings of the IEEE*, 107(8), 1738–1762. 10.1109/jproc.2019.2918951

For Product Safety Concerns and Information please contact our EU
representative GPSR@taylorandfrancis.com
Taylor & Francis Verlag GmbH, Kaufingerstraße 24, 80331 München, Germany